# Technology Research Explained

Ketil Stølen

# Technology
# Research
# Explained

Design of Software, Architectures,
Methods, and Technology in General

 Springer

Ketil Stølen
Jessheim, Norway

ISBN 978-3-031-25816-9      ISBN 978-3-031-25817-6   (eBook)
https://doi.org/10.1007/978-3-031-25817-6

Translation from the Norwegian language edition: "Teknologivitenskap" by Ketil Stølen, © Universitetsforlaget 2019. Published by Universitetsforlaget, Oslo. All Rights Reserved.

This Springer imprint is published by the registered company Springer Nature Switzerland AG
The registered company address is: Gewerbestrasse 11, 6330 Cham, Switzerland

# Preface

Technology science is the science of creating, inventing, or designing new things in the form of human-made objects, so-called artifacts, commonly referred to as technology. In this book, we refer to natural science, social science, and other kinds of science concerned with understanding the world as it is, including the artifacts that already exist, as explanation science.

Technology science often involves explanation science, for example, to better understand the properties of materials needed to build new artifacts. Then the use of explanation science is subordinate to the end goal of inventing an artifact to fulfill some (potential) human needs. Accordingly, explanation science involves technology science to create new specialized instruments, probes, etc. to uncover factual relationships and better understand reality.

The overall goal of this book is to help research practitioners in technology science avoid some of the most common pitfalls or at least make them easier to overcome. The book is for anyone working in technology science, from master's students to researchers and supervisors. In my experience, many technology scientists put too little weight on how they conduct their work. Furthermore, I believe many of these can become significantly better at what they do by being more aware of methodological issues. What I put forth in the form of suggestions, recommendations, and guidelines in this book is what I have found to work for myself, others, and my research projects. Technology science is the intended application domain, but some of it is relevant for science in general.

Ida Solheim helped me carve out early thoughts on technology science, resulting in the technical report [78]. Jannicke Bærheim, Johan Fahlstrøm, Ragnhild Halvorsrud, Amela Karahasanovic, Mass Soldal Lund, Atle Refsdal, Randi Eidsmo Reinertsen, Dumitru Roman, Mariann Sandsund, and Erlend Magnus Viggen have provided comments at some point in the writing process.[1] Dag Frette Langmyhr helped overcome Latex-related obstacles. Thanks to you all.

---

[1] Universitetsforlaget published a version of this manuscript in Norwegian in 2019 [80].

Thanks to fellow lecturers Randi Eidsmo Reinertsen, Mariann Sandsund, Erik Wold, and Laila Økdal Aksetøy. Special thanks to students, course participants, collaborators, and colleagues who (without knowing) have served as "test rabbits" for theories, thoughts, and recommendations presented in the following.

Oslo, Norway                                                                          *Ketil Stølen*
December 2022

# Contents

**1 Introduction** .................................................... 1
   1.1   Technology science in a historical perspective .................. 2
   1.2   Structure ................................................. 3

**2 Technology Science, Explanation Science, and Innovation** .......... 7
   2.1   Knowledge ............................................... 7
   2.2   Technology .............................................. 8
   2.3   Science .................................................. 9
   2.4   Technology science versus explanation science ................. 12
       2.4.1   Technology science ................................. 12
       2.4.2   Explanation science ................................ 14
   2.5   Technology science versus innovation ........................ 15

**3 Technology Science and Its Overall Process** ...................... 17
   3.1   Overall process ........................................... 17
   3.2   The role of reading ........................................ 20
   3.3   The role of writing ........................................ 21
   3.4   Comparison with explanation science ........................ 22
   3.5   Comparison with action research ............................ 25
   3.6   Comparison with technology development ..................... 28
   3.7   Hybrids of different types of research ....................... 29
   3.8   The role of machine learning ................................ 31

**4 Problem Analysis** .............................................. 33
   4.1   Formulation of goals ....................................... 34
   4.2   Characterization of artifact needs ........................... 35
       4.2.1   Identification of artifact needs ...................... 36
       4.2.2   Analysis of artifact needs .......................... 37
       4.2.3   Documentation of artifact needs ..................... 38
   4.3   Mapping of research front .................................. 41

**5  Planning** ............................................................ 43
  5.1  Plan for invention ............................................. 44
      5.1.1  Idea generation ....................................... 44
      5.1.2  Basic techniques for idea generation ................... 45
  5.2  Plan for evaluation .......................................... 50
      5.2.1  Classification of evaluation methods ................... 50
      5.2.2  Method triangulation ................................. 52
      5.2.3  From the general to the special ....................... 53
  5.3  Plan for documentation ....................................... 54
      5.3.1  Invention ............................................ 55
      5.3.2  Evaluation setup and procedures ...................... 55
      5.3.3  Data ................................................. 56
      5.3.4  Materials ............................................ 57
      5.3.5  Interpretation and analysis ........................... 57
      5.3.6  Deductions .......................................... 58

**6  Hypotheses** .......................................................... 59
  6.1  Implicit hypotheses .......................................... 60
  6.2  Working hypotheses .......................................... 61
  6.3  Universal, existential, and statistical hypotheses ................ 64
      6.3.1  Universal hypotheses ................................. 64
      6.3.2  Existential hypotheses ............................... 66
      6.3.3  Statistical hypotheses ................................ 68
      6.3.4  Compound hypotheses ............................... 70
  6.4  Can hypotheses be verified? .................................. 71

**7  Predictions** ......................................................... 75
  7.1  Scientific predictions ......................................... 76
  7.2  Assumptions ................................................ 77
  7.3  Predictions about past events ................................. 79
  7.4  Reusable forms for technology science ........................ 80

**8  Evaluation of Universal Hypotheses** ............................. 83
  8.1  Procedure for evaluating universal hypotheses .................. 84
  8.2  Examples ................................................... 86
      8.2.1  Prediction for prototyping ........................... 86
      8.2.2  Prediction for experimental simulation ................ 87
      8.2.3  Prediction for field experiment ....................... 88
      8.2.4  Prediction for field study ............................ 89
      8.2.5  Prediction for computer simulation .................... 90

8.2.6 Prediction for mathematics ............................ 91
8.2.7 Prediction for logical reasoning ....................... 92
8.2.8 Prediction for survey ................................. 93
8.2.9 Prediction for in-depth interview ..................... 94
8.2.10 Prediction for laboratory experiment ................. 94

9 **Evaluation of Existential Hypotheses** ........................... 97
9.1 Procedure for the evaluation of existential hypotheses ........... 97
9.2 Examples ................................................... 99
9.3 Working hypotheses and evaluation ......................... 101

10 **Evaluation of Statistical Hypotheses** ........................... 105
10.1 Brief introduction to statistical hypothesis testing ............... 105
10.2 Procedure for the evaluation of statistical hypotheses ............ 106
10.3 Examples ................................................. 107
10.4 What if the hypothesis to be evaluated is compound? ............. 111

11 **Quality Assurance** ........................................... 115
11.1 Validity .................................................. 116
11.1.1 External validity ..................................... 116
11.1.2 Internal validity ..................................... 117
11.1.3 Construct validity .................................... 119
11.1.4 Conclusion validity .................................. 121
11.2 Reliability ................................................ 123
11.2.1 Inter-observer reliability ............................. 123
11.2.2 Internal consistency reliability ....................... 124
11.2.3 Parallel-forms reliability ............................. 125
11.2.4 Test-retest reliability ................................ 126

12 **Publishing** ................................................. 129
12.1 Selection of publication channel ............................. 129
12.1.1 Scientific poster ..................................... 130
12.1.2 Scientific abstract ................................... 130
12.1.3 Popular scientific publication ......................... 131
12.1.4 Scientific article ..................................... 131
12.1.5 Scientific report ..................................... 132
12.1.6 Master's thesis ...................................... 133
12.1.7 Doctoral thesis ...................................... 133
12.1.8 Scientific book ...................................... 134
12.1.9 Patent .............................................. 134
12.2 Reuse ................................................... 134

**13   Article Writing** ................................................... 137
  13.1  Structure ..................................................... 137
  13.2  Introductory part ............................................. 139
      13.2.1  Title .................................................. 139
      13.2.2  Author list ............................................ 140
      13.2.3  Abstract ............................................... 141
      13.2.4  Keywords ............................................... 141
      13.2.5  Introduction ........................................... 141
      13.2.6  Characterizing artifact needs .......................... 142
  13.3  Research method part .......................................... 142
  13.4  Artifact part ................................................. 143
  13.5  Evaluation part ............................................... 144
  13.6  Discussion part ............................................... 144
      13.6.1  Discussion of evaluation results ....................... 144
      13.6.2  Discussion of validity and reliability ................. 145
      13.6.3  Discussion of whether artifact needs are satisfied ..... 146
      13.6.4  Discussion of related work ............................. 146
  13.7  Closing part .................................................. 146
      13.7.1  Conclusion ............................................. 146
      13.7.2  Further work ........................................... 147
      13.7.3  Thanks ................................................. 147
      13.7.4  Bibliography ........................................... 147
      13.7.5  Attachments ............................................ 148
  13.8  If you get stuck .............................................. 148
      13.8.1  Getting started ........................................ 148
      13.8.2  Establishing a common thread ........................... 149
      13.8.3  Little to discuss ...................................... 150
      13.8.4  Conclusion says nothing ................................ 151
      13.8.5  Nothing more to cut .................................... 151
      13.8.6  Artifact needs do not fit in ........................... 152
      13.8.7  Hypothesis does not fit or is missing .................. 152
      13.8.8  Are predictions needed? ................................ 153

**14   Technology Science from the Perspective of Philosophy of Science** ... 155
  14.1  Main directions of the philosophy of science ................. 155
      14.1.1  Empiricism ............................................. 155
      14.1.2  Inductionism ........................................... 156
      14.1.3  Positivism ............................................. 156
      14.1.4  Logical empiricism ..................................... 157
      14.1.5  Falsificationism ....................................... 158
      14.1.6  Paradigm thinking ...................................... 159
      14.1.7  Epistemological anarchism .............................. 161
      14.1.8  Probabilism ............................................ 161
      14.1.9  Experimentalism ........................................ 162

14.2 Technology science in this picture............................ 162
    14.2.1 Technology science and empiricism..................... 163
    14.2.2 Technology science and inductionism .................. 163
    14.2.3 Technology science and positivism .................... 164
    14.2.4 Technology science and logical empiricism ............. 165
    14.2.5 Technology science and falsificationism ................ 165
    14.2.6 Technology science and paradigm thinking.............. 166
    14.2.7 Technology science and epistemological anarchism........ 166
    14.2.8 Technology science and probabilism ................... 167
    14.2.9 Technology science and experimentalism ............... 167

**Overview of Definitions** ........................................... 169

**References** ...................................................... 175

**Index** ......................................................... 179

# Chapter 1
# Introduction

Technology science is about generating knowledge in the form of new artifacts, meaning they did not exist before the research was initiated. On the other hand, explanation science provides new knowledge about the world, including the already existing technology and artifacts.

This book focuses on technology science in general and its overall process. But what does "overall process" mean? Is the overall process the same as the research method? Both yes and no. The phrase "overall process" of technology science denotes a general research approach – the procedures or methodological aspects constituting best practices common to the different disciplines of technology science.

What constitutes best practice? It is, obviously, a relevant question. Nonetheless, in many fields of technology science, it does not receive the attention it deserves. I did technology science for almost ten years before I began to think much about what was good practice for my work. When I look back, it is not difficult to identify potential for improvement. I also see this potential for improvement in others, including past and present colleagues, project partners, as a peer reviewer, and in my role as periodical editor.

When I ask inexperienced researchers, such as a master's student or a doctoral fellow, how they plan to go forward to carry out their research project, I often get an answer along the following lines:

> First, I have to read up on the topic. Then I define the research problem or identify research questions, do the actual research, and finally, write it up in a report, dissertation, or article.

My usual response is that you may succeed using this strategy but that it is not recommendable in the general case. In my experience, it may easily cause delays and reduce the quality of results. First, it is risky to read a lot before defining the problem to be addressed. There are incredible amounts of literature available on most topics, and the danger of reading much of what is not helpful and little of what you need is very real. Second, if you start writing up only after completing the research, you will often find that you do not have the results, data, or documentation needed.

© The Author(s), under exclusive license to Springer Nature Switzerland AG 2023
K. Stølen, *Technology Research Explained*, https://doi.org/10.1007/978-3-031-25817-6_1

There are many specialized research methods, and myriads of books and articles cover them, often thoroughly and educationally. Examples are research methods for hypothesis testing and statistical analysis [5], for different types of experiment setup [74], [99], [48], qualitative research [17], [15], and creativity in research [62]. Much of this literature addresses natural sciences such as physics or chemistry, social sciences like sociology or economics, or focuses on medicine.

This book differs from those in two main respects: First, by focusing on creating, producing, or inventing new artifacts – in other words, technology science; Second, by describing a general approach to technology science linking together specialized research methods.

## 1.1 Technology science in a historical perspective

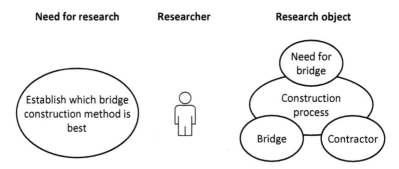

**Fig. 1.1** Science about design: Study of methods for bridge construction.

Technology science has its historical roots in the 1960s [18]. The book, *The Science of the Artificial* [76], published by Herbert Simon (1916–2001) in 1969, introduced the notion of design science. Hevner and Chatterjee [38] distinguish between *researching design* in the meaning of "science of design" and *design as research* in the sense of "design as science."

In the first sense, the science of design, design science revolves around design and design processes in general, regardless of whether the design represents research or not. The research involves studying designers in their work to establish knowledge of existing design processes and design methodology. A researcher may, for example, try to determine if a particular process or method is more efficient or provides better quality than another. As indicated in Figure 1.1, the research may evaluate existing bridge construction methods. The goal may be to identify which is best suited concerning one or several qualities, such as cost-effectiveness or safety. One possible approach is to study documentation from already completed bridge construction projects. In each project, there is a need for a new bridge (which must not be confused with the research need). Furthermore, there are one or more contractors, a construction process, and the outcome is a new bridge. A research result may be that one method for bridge construction is more cost-effective than another, given several more specific assumptions and requirements.

The second meaning, design as science, corresponds to what we in this book call technology science. A technology scientist invents and delivers results in the form of new artifacts. Figure 1.2 describes the relationship between explanation science, technology science, and the two meanings of design science.

$$\text{science of design} \subset \text{explanation science}$$
$$\text{design as science} = \text{technology science}$$

**Fig. 1.2** Technology science and the two forms of design science.

Science of design is about understanding design and design processes and therefore belongs to explanation science, which motivates the subset operator "$\subset$". After all, explanation science is far more than the science of design. Design as science coincides with what we call technology science, which explains the equality symbol. Science of design addresses important unresolved issues uniquely or innovatively. Alternatively, it solves problems in a better or more economical manner. Moreover, the main difference between routine design and design as science is an identifiable knowledge contribution [39]. Technology science and its methodological aspects have attracted much interest in line with the explosive growth in new IT-based technology. See [88] and [21] for alternative presentations of technology science.

## 1.2 Structure

This book consists of 14 chapters. Following this first, introductory, chapter are two chapters providing the foundation for the rest of the book. These chapters clarify the meaning of key concepts and describe an overall process for technology science. Chapters 4–11 are about this process. Chapters 4, 5, and 6 concern problem analysis, research planning, and hypothesis formulation. The following five chapters aim at evaluation. Chapter 7 introduces the concept of prediction, which plays a fundamental role in evaluating hypotheses. Chapters 8–10 address the evaluation of universal, existential, and statistical hypotheses. Chapter 11 concerns quality assurance and introduces the concepts of validity and reliability. Then, in Chapters 12 and 13, we address publishing, with an emphasis on the specifics of technology science. The last chapter, Chapter 14, briefly introduces the philosophy of science. It explains how technology science fits into this picture. A more detailed summary of each chapter follows below.

- *Chapter 2 – Technology Science, Explanation Science, and Innovation:* To communicate well, we depend on a neat conceptual vocabulary that others understand as we do. The goal of Chapter 2 is to establish such a vocabulary for this book. The chapter introduces concepts such as knowledge, technology, and science. It also explains the relationship between technology science, explanation science, and innovation.
- *Chapter 3 – Technology Science and Its Overall Procedure:* We need an overall process to present the different aspects of technology science. This chapter introduces such a process. The chapter distinguishes between three main phases: the needs characterization phase, the invention phase, and the evaluation phase. Having introduced these phases, we then describe the roles of reading and writing in the context of this overall process. We then look at how this technology science process differs from other kinds of science, such as explanation science. We also relate action research commonly employed to study social structures, processes, and businesses to technology science. Then, we draw the borderline between technology science and technology development and highlight that many research projects in technology science involve both research and development. Finally, given its importance as a research method, we position machine learning in the picture. We consider its impact on the research process and to what extent it alters the role of the researcher.
- *Chapter 4 – Problem Analysis:* This chapter concerns capturing and describing the research problem. It involves formulating research goals, characterizing the artifact needs, and establishing an overview of available technology and relevant knowledge.
- *Chapter 5 – Planning:* Once the artifact needs are understood and properly documented, we must establish a plan for conducting the research. The chapter distinguishes between the plans for the invention phase, the evaluation phase, and the documentation.
- *Chapter 6 – Hypotheses:* This chapter is concerned with the notion of hypothesis. There are traditions and fields in technology science where hypotheses seldom are formulated explicitly. We, therefore, start by shedding light on the implicit role of hypotheses in such domains. We then address the so-called working hypotheses and their role in the invention phase. We also classify hypotheses according to whether they are universal, existential, and/or statistical. The chapter ends with a discussion of the extent to which hypotheses can be falsified or verified.
- *Chapter 7 – Predictions:* A researcher must be able to argue for his or her findings. This requires some examination or evaluation of which predictions are essential. In this chapter, we first look into the notion of scientific prediction and the role of predictions in the evaluation phase. Then we explain the relationship between predictions, assumptions, and facts. A prediction depends almost always on existing theory and knowledge. We also emphasize that predictions can be about past events as long as the concrete tests, experiments, etc. it refers to will

be carried out in the future. Finally, we introduce ten reusable forms or patterns for formulating predictions.

- *Chapter 8 – Evaluation of Universal Hypotheses:* In general, the procedure for hypothesis evaluation depends on the structure of the hypothesis. This chapter is about evaluating universal hypotheses. We put particular emphasis on the use of reusable forms or patterns.
- *Chapter 9 – Evaluation of Existential Hypotheses:* This chapter addresses the evaluation of existential hypotheses. As in the previous chapter, we use patterns to formulate hypotheses and predictions. Towards the end, we address evaluating working hypotheses refined or detailed into new hypotheses, based on the evaluation outcome for the latter.
- *Chapter 10 – Evaluation of Statistical Hypotheses:* This chapter concerns evaluating statistical hypotheses. Initially, we give a brief introduction to statistical hypothesis testing. As in the two previous chapters, we make use of patterns. The chapter ends with a section on the evaluation of compound hypotheses.
- *Chapter 11 – Quality Assurance:* Mistakes may happen in an evaluation process, and it is essential to check and justify that the evaluation is solid and correct. Quality assurance involves checking to what extent the requirements for validity and reliability are satisfied. This chapter presents these requirements and their relationships. The chapter provides insights into what we need to look out for to ensure quality.
- *Chapter 12 – Publishing:* Publishing is a duty for any research scientist. This chapter addresses publishing in technology science. We distinguish between nine different kinds of publications. We address issues related to reuse and so-called self-plagiarism.
- *Chapter 13 – Article Writing:* The article format is the gold standard for scientific publishing. We start by giving recommendations on the structuring: Which sections to include and in what order. We then go through these sections one by one and address, in particular, those aspects that are specific to technology science. Anyone who has tried to write a scientific article knows it can be difficult. The chapter dedicates the last part to common challenges.
- *Chapter 14 – Technology Science from a Philosophy of Science Perspective:* This chapter consists of two parts. The first part briefly overviews the relevant philosophical directions, emphasizing the last 100 years. The other part then positions technology science as presented in this book in the landscape described in the first part.

To make it easier to maintain an overview, Appendix A summarizes all definitions. The book ends with a bibliography and an index. The index employs the following convention:

- Page number in bold refers to where the term is defined (assuming there is one).
- Page number in italic refers to where in Appendix A the definition is repeated.

# Chapter 2
# Technology Science, Explanation Science, and Innovation

Concepts such as science, research, technology, and innovation are difficult to relate. One problem is their abstract nature; another is that their interpretation varies depending on the context and the field. This chapter will explain how this book interprets these and related concepts.

## 2.1 Knowledge

We live in a reality that exists independently of ourselves. When we die, this reality still exists. Most of us take this for granted.

**Definition 2.1** *Reality* is the outer world. What we face in our natural attitude to the outside world, and what we can indirectly deduce about it.

Some philosophers question the existence of a reality beyond what is directly observable. However, we will not spend time on such views in this book. For most of us, the men and women of science included, the existence of a reality independent of (but not unaffected by) ourselves is obvious and something on which we base our thinking and daily work.

We can observe some of this reality directly. We can only indirectly deduce other parts, like the genetic code, light bending, and electrons' behavior. What we know about reality, we call knowledge, and when we learn something new, we expand or revise this knowledge.

**Definition 2.2** *Knowledge* is what is learned or understood, what we know.

Knowledge is about reality. This reality includes other human beings. You, a living and thinking organism, are part of this reality. Since you are part of reality, your knowledge is also part of reality. In this book, however, we sharply distinguish

K. Stølen, *Technology Research Explained*, https://doi.org/10.1007/978-3-031-25817-6_2

between knowledge and what this knowledge is about, namely, reality. This is a simplification or an abstraction, but a common and useful one.

We are constantly trying to gain more knowledge. To this end, we use our senses supported by various aids. Some of these are human-made. Others are natural. If we aim to measure the depth of a puddle, we can use a stick that we find on the ground – in other words, a natural tool. A telescope is a human-made aid to study the lunar surface. Cognition is often referred to as a human's endeavor to acquire knowledge of reality.

> **Definition 2.3** *Cognition* is the activity of perceiving something as it is (in reality), regardless of the cognitive subject.

Knowledge should be objective. It should not depend on who observes or conducts the cognition. The latter is an ideal that we constantly strive towards without fully being able to reach. Research on social processes and social structures is often challenging in this respect. Objectivity is not easy in other disciplines either. In quantum physics, the problem is at the forefront: It is theoretically impossible at a given moment to accurately measure both the position as well as the velocity of an electron. The problem is that the measurement interferes with the electron's behavior. The phenomenon is known as Heisenberg's principle of uncertainty [82]. The more accurately we measure the one, the more uncertain becomes the other.

## 2.2 Technology

In this book, we are concerned with things and objects made by man. These we refer to as artifacts.[1] The term artifact has its origins in the Latin term *arte*, which means "(trained) ability" [94].

> **Definition 2.4** An *artifact* is a thing, an object, or a phenomenon created by humans.

The term artifact defined above covers everything from simple tools such as a flint ax, knife, or slingshot to advanced technology such as a weather satellite. Artifacts can also be immaterial. A business process is an artifact; so is a computer program.

Technology has many different definitions, but they are all spun around "the practical application of knowledge to make things." The term technology originates from the Greek *technologia* and translates into "systematic use of tekhne", where *tekhne* means "art or skill" [58]. Here the focus is on the method and not the artifacts pre-

---

[1] In a laboratory, an artifact commonly denotes a human-made mistake or deviation. This interpretation must not be confused with our general definition; it is one specific case.

pared using this method. Today, the term technology has a broader meaning. The American sociologist Read Bain (1892–1980) [3] has inspired the definition below.

**Definition 2.5** *Technology* includes all human-made objects and the skills we use to manufacture and employ them.

Tools, machines, weapons, instruments, rooms, clothes, medicines, and means for communication and transport are things or objects. Methods, processes, and social structures manufacture and use these. All these things are man-made and therefore artifacts.[2]

## 2.3 Science

Science has its roots in the philosophy of antiquity. Science in antiquity was first and foremost natural science. The aim was to understand the world as it appeared. Science concerns (or is about) knowledge in which we have high trust. That is, we acquire knowledge of nature, space, the human body, etc. Today the concept of science is more comprehensive, and there are far more sciences. Information science and media science are examples of relatively new ones.

**Definition 2.6** *Science* is a systematically arranged body of methodologically established knowledge.

Methodologically established knowledge refers to knowledge generation – the process of cognition. In science, we refer to this process as research. According to Merriam-Webster [57], research is

> investigation or experimentation aimed at the discovery and interpretation of facts, revision of accepted theories or laws in the light of new facts, or practical application of such new or revised theories or laws.

Slightly more abstract, we define research as follows.

**Definition 2.7** *Research* is a systematic process for generating new knowledge.

A researcher is a person doing research. There are many ways to do research. These various approaches are known as research methods.

**Definition 2.8** A *research method* is a specialized approach or procedure to conduct research.

---

[2] In this book, technology and artifact have the same meaning. However, technology is a "charged" term in English. Few think of a procedure for extinguishing wildfires or medicine for treating infections as technology. Therefore, we use the artifact concept more often since its meaning in English is more open.

It is common to distinguish between basic research and applied research. Basic research is about gathering knowledge independently of whether this knowledge has practical utility or not.

**Definition 2.9** *Basic research* is research aimed at satisfying the need to know.

The term applied research has several interpretations. That applied research is practical and is about solving practical problems is generally accepted. One interpretation limits applied research to using results from basic research to solve practical problems. Research addressing practical issues, in general, is another interpretation. In the latter case, applied is interpreted as "useful" The definition below is of the second kind.

**Definition 2.10** *Applied research* is research to find solutions to a practical human problem.

An example of applied research founded on basic research is Bragg's method to locate the position of enemy artillery cannons by exploiting the theory of sound waves.

*Example 2.1 (Locating enemy artillery cannons).* William Lawrence Bragg (1890–1971), who in 1915 shared the Nobel Prize in physics with his father, served on the British side during World War I. Bragg was tasked [46] with inventing a method to deduce the position of enemy artillery cannons from the pressure waves that arose when they fired. A problem encountered by Bragg was that cannons produce very low-frequency sound waves that are difficult to detect using the equipment of the time. This frustrated Bragg because there is a lot of energy in the sound waves generated by a cannon shot. The latrine used by Bragg while stationed in Flanders was in a windowless enclosure. When the door was closed, the only opening to the outside was the underside of the seat. Every time a British "6-inch" was fired at a distance of 400 meters, Bragg lifted from the seat independent of whether he could hear the bang.

Another problem was that the bang when the projectile broke the sound barrier drowned out the bang from the cannon firing. Inspired by one of his corporals, William Sansome Tucker (1877–1955), Bragg devised a solution for reading the sound waves based on existing apparatus for measuring wind speed. These so-called Tucker microphones were positioned in the terrain. By measuring the time difference, it was possible to calculate the positions of the cannons with sufficient accuracy. The method made use of results from basic research, such as knowledge of the speed of sound and how it varies depending on wind and temperature.

An example of applied research not based on basic research is Vass-Per's leveling tool to determine ground inclination.

*Example 2.2 (Leveling tool for building waterways for irrigation).* Peder Pedersen Dagsgardødegård (1782–1846) from Skjåk in Norway, better known as Water-Per

("Vass-Per" in Norwegian), was an expert in calculating and building waterways for irrigation [53]. Skjåk is surrounded by high mountains and is very dry. Skjåk gets, on average, only 278 mm of rain per year, not much more than in many deserts.

Water-Per was the designer of many waterways but is primarily known for the Bordvassvegen waterway. This waterway starts at 1450 meters and ends at 450 meters above sea level. With the technology of that time, this was a very demanding task. He had to ensure that the farmers got enough water during the whole season but not so much that the fields at any point flooded. To assist in this work, Water-Per invented a leveling tool to determine the slope of a hill. Water-Per did not build on results from basic research but did applied research in our meaning of the term.

Definition 2.6 of science requires knowledge to be systematically arranged and established methodologically. Hence, science is characterized by regularity and structure. Knowledge in science has an important component called a theory.

**Definition 2.11** *A* theory is a system of (partially) confirmed statements that determine or explain the relations between phenomena.

Not all knowledge is theory. That the Swedish King Karl XII (1682–1718) was killed at Fredriksten fortress in Halden close to the Norwegian border is knowledge of reality but not theory because it does not explain anything. It represents a fact in the sense: Something that has happened.

**Definition 2.12**  A *fact* is a true statement about some past event.

Like most people, I consider Karl XII's death at the Fredriksten fortress as a fact.[3] I am also sure that I sat on a red couch at a hotel reception in Malaga and wrote the first draft of this section. Hence, I consider that also to be a fact.

Philosophers like to point out that we cannot be sure about anything. The lighting at the front desk, for example, may have led me to believe that the couch was red, or the coffee I drank may have contained a drug that brought my senses into disarray. Furthermore, I am more confident that the couch was red than about Karl XII's death at Fredriksten. After all, the last event happened almost 250 years before I was born. In practice, however, such considerations are not very fruitful. We have so much confidence in certain past events that we classify them as facts, which is the line we follow in this book.

The initiating cause or starting point for a new theory is typically a set of questions and some initial guesses as to what the answers may be. An educated guess is a guess based on knowledge [11]. Educated guesses are called hypotheses.

**Definition 2.13**  A *hypothesis* is an educated guess expressed as an assertion.

---

[3] That Karl XII died at Fredriksten's fortress is not disputed by anyone. Whether he was shot by a Swedish or Norwegian soldier and by whom has been debated for centuries [52].

The researcher may postulate several alternative hypotheses. Their evaluation involves different kinds of examinations or analyses.

Most hypotheses are in a short while rejected. But some hypotheses survive thorough evaluations. Then they are no longer called hypotheses even though they are nothing but hypotheses whose correctness we have great faith in. Instead, they belong to what we call knowledge or theory.

Empirical research is research based on experience, observation, etc. This book is almost exclusively about empirical research and empirical hypotheses.

**Definition 2.14** An *empirical hypothesis* is a hypothesis about reality.

Non-empirical hypotheses can be evaluated, proved, or disproved, purely mathematically or logically at an abstract level, independent of reality. Mathematics and logic are of importance also when evaluating empirical hypotheses, but then always in combination with other evaluation methods.

## 2.4 Technology science versus explanation science

In this book, as explained earlier, we distinguish between two main kinds of science, namely technology science and explanation science, respectively. Explanation science is concerned with describing, mapping, and understanding reality. Technology science is about expanding this reality with new artifacts – that is, new technology. Most sciences, like physics, chemistry, information science, and sociology, include both kinds. We will further clarify the relationship between technology science and explanation science in the following.

### 2.4.1 Technology science

The engineering sciences are at the core of technology science, but the concept goes beyond that. Today's chemists are constantly inventing new artifacts in the form of substances and materials, such as medicines, while geneticists design artifacts in the form of new or refined species. Although chemists and biologists in the past mainly did explanation science, they also invented artifacts such as instruments, new substances, methods, and procedures.

**Definition 2.15** *Technology science* is science aiming at expanding reality with new or significantly better artifacts.

A technology scientist endeavors to create or invent new artifacts. That is, artifacts that do not yet exist. An extreme example is a vessel capable of traveling to a distant galaxy (Figure 2.1). Achieving this is not realistic today, but we have in the past succeeded in bringing about other kinds of technology that some centuries earlier seemed utopian, not least the invention of the powered aircraft.

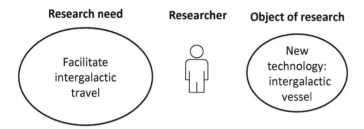

**Fig. 2.1** Technology science: Invention of intergalactic vessels.

*Example 2.3 (The powered aircraft).* In 1799, George Cayley (1773–1857) defined the concept of the modern airplane, consisting of fixed wings and separate lifting, propulsion, and control systems [89]. However, it took more than one century, until 1903, before the brothers Orville Wright (1871–1948) and Wilbur Wright (1867–1912) succeeded in performing the first controlled flight with a powered aircraft. It is undoubtedly one of the most outstanding achievements of technology science ever. Although the Wright brothers took the last step, they built on the results and experiences of countless others. The Wright brothers' flight represents something typical of technology science, namely proving the correctness of a principle by building a prototype or specimen that works.

The airplane exemplifies success, but technology science may, of course, also fail.

*Example 2.4 (The dream of making gold).* An example of failure is the alchemists' attempts to make gold from simple or cheap ingredients. Gold itself is not an artifact but a natural element. It was not gold itself the alchemists were trying to design, but an artifact in the form of a procedure or method for making gold cheaply. Today gold may be produced in a laboratory, but only utilizing a particle accelerator or nuclear reactor [92]. The cost associated with the manufacture is many times the market price of gold.

## 2.4.2 Explanation science

Sciences such as physics, chemistry, and biology belong at least historically mainly to explanation science, although they have also made outstanding contributions to technology science. Most social sciences also belong mainly to explanation science.

**Definition 2.16** *Explanation science* is science aiming at understanding reality as it is.

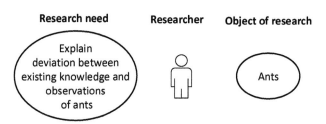

**Fig. 2.2** Explanation science: Study of ants.

Explanation science focuses on capturing, understanding, and characterizing reality. Explanation science studies reality as it is, in contrast to the technology science that extends reality with new artifacts.

In explanation science, the starting point is typically a deviation between knowledge and theory on the one hand and what is observed on the other. As suggested in Figure 2.2, there may be observations of ants incompatible with what we already "know" about ants. A researcher may study these deviations and try to establish new explanations that deepen or enhance existing knowledge. A famous example of such a discrepancy between observation and theory gave rise to the hypothesized existence of the planet Vulcan [49].

*Example 2.5 (The dark planet Vulcan).* In 1859 the French mathematician Urbain Le Verrier (1811–1877) published a detailed study of Mercury's orbit that showed a discrepancy between the observed trajectory and what it should be according to Isaac Newton's (1643–1727) theory of gravity. In an attempt to explain this discrepancy, Le Verrier hypothesized that there exists an unknown planet, Vulcan, whose orbit around the Sun is inside that of Mercury. Many astronomers searched intensely for this planet without success. In the absence of convincing observations, Vulcan was referred to as the dark planet.

Later, in 1919, it became clear that the phenomenon could be explained by Albert Einstein's (1879–1955) general theory of relativity. It is usually considered[4] the first empirical confirmation of the general theory of relativity.

---

[4] Here, as so often, there are different views – see, for example, [43].

Even today, astronomers are looking for dark phenomena. According to current theory, 23% of the total energy of the Universe is represented by dark matter, and 73% is dark energy. Whether these phenomena exist or are just deductions from erroneous theories is disputed even among leading physicists [69].

## 2.5 Technology science versus innovation

While technology science is all about creating new artifacts, innovation, as the term is commonly used today, focuses on the practical exploitation and commercialization of new ideas. Such an idea can be a combination of old ideas, a paradigm that violates common sense, a formula, or a particular scheme perceived as new by those involved – the people or users for whom it is intended [86]. As long as the idea is perceived as new by those involved, it is innovative, even though others perceive it as a copy or imitation of something that already exists elsewhere.

An idea leading to innovation may originate from science, but often it does not. Innovation may, for example, be user-driven and appear as the outcome of collaboration between users and suppliers or when businesses actively collect and analyze user requirements. Innovation is also often cost-driven. The focus of innovation is then to reduce costs in manufacturing products and services.

There are many definitions of innovation. In 1934 the economist and social scientist Joseph Schumpeter (1883–1950) defined innovation as new combinations of new or existing knowledge, resources, equipment, and so on [73], [24]. In a very general sense, innovation may be defined as follows [19].

**Definition 2.17** *Innovation* is the production or adoption, assimilation, and exploitation of a value-added novelty in economic and social spheres; renewal and enlargement of products, services, or markets; development of new methods of production; or the establishment of new management systems. It is both a process and an outcome.

Although innovation originates from *innovare* in Latin, which means to create something new, the concept as used today denotes the practical use or the commercialization of something new in a broad sense. It may be a new commodity, a new service, a new production process, or an organizational structure launched in the market or implemented to foster financial values. A new idea or invention does not become an innovation until it has matured into a practical application. Often, it is not the creator of an idea who does the innovation. The innovation may occur long after the idea was conceived and in a different location.

There are many forms of innovation. The following classification, based on [37], distinguishes between four categories:

- *Incremental innovation:* Refines and/or extends an existing design or architecture.

- *Modular innovation:* Replaces one or more components of existing architecture with new ones, such as an analog phone with a digital one.
- *Architectural innovation:* Ties together or combines existing components in a new way.
- *Radical innovation:* Establishes a new dominant concept based on new components and architecture.

The transition between these categories is gradual. Some innovations belong to several of these categories or are located on the border between two or more.

Innovation has different driving forces. Research-driven innovation is particularly relevant for this book. Here is a famous example of just that.

*Example 2.6 (Thomas Alva Edison and the light bulb).* Many probably think first and foremost of Thomas Alva Edison (1847–1931) as a great inventor, and not least as the inventor of the light bulb. Edison was undoubtedly a great technology scientist, though he did not invent the light bulb and spent a significant part of his career fostering innovation. Edison invented the first commercially practical, white glowing light bulb in 1879. The light bulb as such was conceived even before Edison was born, and at least 22 variants of white incandescent light bulbs existed before Edison launched his version in 1879 [28].

The remarkable thing about Edison's light bulb was that it could be taken into commercial use. He even provided commercialization by establishing the necessary infrastructure as a power station and distribution apparatus. In 1880 Edison started the Edison Illuminating Company. Two years later, the company's power station in Pearl Street in New York entered into use. Its distribution system provided 110 volts of direct current to 59 customers in Lower Manhattan.

# Chapter 3
# Technology Science and Its Overall Process

There are many research practices and working styles. Not all are equally effective, and some are not recommendable. In this chapter, we describe the overall process of technology science on which we base the rest of the book. This process is later refined and specialized. Reading and writing are essential ingredients in all scientific research. This chapter outlines the roles of reading and writing in the above-mentioned process.

This chapter also characterizes the relationship between the process of technology science on the one hand and the processes of explanation science, action research, and technology development on the other.

Machine learning has become a popular tool in all kinds of research. Some argue it is a game-changer that alters the scientific process. We end this chapter by discussing the relevance of such claims.

## 3.1 Overall process

Technology science aims at inventing new or improved artifacts. It may be a new robot design, a new material, a new design principle for high-speed motorways, a new recipe for cough syrup, or a new or improved program for patient care.

A technology scientist will rarely deliver a fully finished commercial product. The outcome is typically a design or prototype that others may take further, refine or develop for business or societal purposes.

The starting point for a technology scientist is some unfulfilled (potential) human needs that a new artifact may fulfill. We have written *(potential)* because the needs could be futuristic, meaning they do not exist in the current context. If, for example, flying vehicles replace today's passenger cars, there will be new needs for traffic control.

The first challenge is to identify and characterize these needs. The needs capture may involve existing users (if the challenge is to improve some existing artifact)

K. Stølen, *Technology Research Explained*, https://doi.org/10.1007/978-3-031-25817-6_3

or new or potential users (if the kind of artifact in question does not yet exist). The views and opinions of other stakeholders, such as funding or government authorities, must also be accounted for.

Often, the needs are not identified by the researcher but rather by some other stakeholder. This may be a business, a group of individuals (for example, people with a particular disorder), or a public institution. The task of the researcher is to create, invent, or enhance an artifact that has the potential to help fulfill these needs.

Once the needs are characterized, the invention phase may start. Hopefully, the researcher has a good idea or hypothesis for how to proceed or at least a strong belief that the needs can be fulfilled.

If the research is fertile and results in some sort of invention, it must be evaluated or tested for the needs that were the starting point. The researcher may claim to have succeeded if the evaluation outcome is positive. If not, and this is not due to evaluation errors, we must either improve the artifact design or adjust the needs to simplify the challenge, for example, by only addressing some of the needs.

The description above gives the impression that individuals conduct technology science. This is usually not the case, but a simplification for presentation purposes that we will make use of also later in the book. Research projects are usually joint efforts of groups of researchers. Different members of the group may have specialized tasks reflecting their competencies. For example, a researcher capturing and analyzing the needs may not be involved in the evaluation.

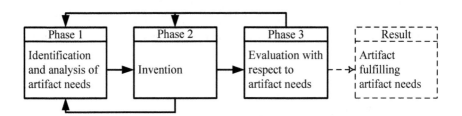

**Fig. 3.1** Overall process of technology science.

The overall process of technology science consists of three phases (see Figure 3.1):

- Phase 1 – *Identification and analysis of artifact needs:* The aim is to characterize the needs for the new or improved artifact. Initially, it may be unclear what these needs are. There may be several stakeholders and conflicting views. It can be demanding to capture, clarify, and document the needs appropriately.
- Phase 2 – *Invention:* The aim is to invent, design, or create an artifact that satisfies the needs. This is the most creative stage. Hypotheses are formulated, and new ideas are hatched. Sometimes the invention is modest, though useful, such as adapting an existing artifact to cover new needs. At other times, the invention can be groundbreaking.

- Phase 3 – *Evaluation with respect to artifact needs:* The aim is to decide whether or to what extent the new artifact fulfills the identified needs. This may involve experiments and various kinds of investigations.

For more clarity on what the three phases described above mean, let us look at an example from history, namely the invention of the telescope. It is a mythic event. Our presentation is based on [32].

*Example 3.1 (The telescope).* The telescope was reinvented[1] by the Dutch eyewear maker Hans Lipperhey (1570–1619) in 1608. Galileo Galilei (1564–1642) heard a rumor about this instrument in Venice in 1609 and saw its commercial potential. When he was told that the Dutchman was on his way to Venice to sell the instrument, he built a significantly better version in 24 hours. Galilei's new design involved two lenses and a cylinder, nothing more. Lipperhey used two concave lenses with the result that the observed phenomenon was depicted upside down. Galilei's telescope, on the other hand, did not suffer from this problem because he used a concave and a convex lens. Galilei refined his version to enlarge ten times, while Lipperhey's telescope only managed three.

How should Galilei's contribution be understood concerning the three phases above? For simplicity's sake, let us assume that Galilei was aware that Lipperhey's telescope depicted upside down and magnified three times only. The artifact need can then be expressed as follows:

An instrument that magnifies at least four times and depicts the right way up.

That the instrument (1) *magnifies at least four times* and (2) *depicts the right way up*, we can think of as success criteria that Galilei aimed to fulfill. Nobody knows how Galilei thought, but concerning success criterion (2), he could have formulated the following hypothesis:

By using a pair of concave and convex lenses instead of a pair of concave lenses, it is possible to make a variant of Lipperhey's telescope that depicts the right way up.

This is an example of an *existential hypothesis*. It claims it is possible to make at least one copy of something. It does not say that any variant of Lipperhey's telescope based on a pair of concave and convex lenses will depict correctly, but only that at least one will fulfill the need. Hence, it claims at least *one* variant exists, which is why it is referred to as existential. In technology science, the evaluation of an existential hypothesis consists in building a kind of prototype that satisfies the hypothesis, and, this was precisely what Galileo did.

---

[1] The English astronomer Leonard Digges (ca. 1515 – ca. 1559) is considered the first inventor of the telescope. This happened more than 50 years before Lipperhey and Galilei invented their versions (but probably neither Lipperhey nor Galilei knew Digges's telescope).

## 3.2 The role of reading

A researcher must be up to date in their field of research. This means having an overview of already existing knowledge in the relevant domain of research. It does not imply that a young researcher must spend years reading (after graduation) before even thinking of doing research. First, researchers usually work in groups where some are experienced and know the literature. Second, if the field is narrow, it is less demanding to obtain an overview. Most young researchers work in very specialized areas, to begin with. Figure 3.2 characterizes the role of reading in the overall process presented in Figure 3.1.

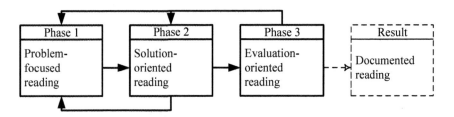

**Fig. 3.2** The role of reading in the overall process.

- *Problem-focused reading:* In technology science, problem-focused reading addresses two main issues:
  1. Acquire knowledge about the needs (for example, published results on the training needs of stroke patients if the task is to make a training device for such).
  2. Gain knowledge of the suitability of existing technology (if some training needs of stroke patients are satisfactorily met by existing technology, the problem can be simplified or refined).

  The problem is usually not fully understood when the research starts, and problem-focused reading is performed as part of the problem analysis. Initially, we typically have only a coarse-grained sketch or high-level understanding of the relevant issues. Based on this rough sketch, the direction of the introductory reading is determined. Hopefully, it leads to a deeper understanding and a more precise problem characterization. It is used to refine the reading further, and this way, we proceed until the problem is specified.
- *Solution-oriented reading:* This is a more specialized and profound form of reading. Now, there is not necessarily much literature to choose from. It is more a matter of getting a good overview of the existing options and the extent to which they are appropriate in our context.

- *Evaluation-oriented reading:* Evaluation-oriented reading is usually quite to the point. The issue is acquiring knowledge on the best ways to evaluate or test that the artifact satisfies the identified needs.
- *Documented reading:* When we write an article or report presenting our findings, we need to characterize what our new artifact does better than already existing technology and provide evidence to that extent. To convince readers that we have not forgotten or ignored significant competing contributions, we may document the principles according to which they have been identified and selected. The documentation may, for example, name the search engines or databases we have used, with which keywords, and when.

## 3.3 The role of writing

Documenting decisions, ideas, developments, etc., is essential throughout the research process. Written wording clarifies thoughts and promotes effectiveness and efficiency. Written text and drawings (possibly supported by speech) may be preferable to speech even when the researchers are within speaking distance.

Figure 3.3 characterizes the role of writing in the overall process presented in Figure 3.1.

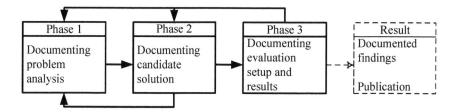

**Fig. 3.3** The role of writing in the overall process.

- *Documenting problem analysis:* Phase 1 involves performing a problem analysis. The results of this analysis must be documented. In particular, the identified needs must be specified.
- *Documenting candidate solution:* Ideas and understanding will evolve and be refined during the invention phase. If the research is fertile, we end up with a hypothesis referring to a detailed description of the artifact constituting a potential solution.
- *Documenting evaluation setup and results:* Accurate and proper documentation is essential to allow others to check and test the correctness of our evaluation.

- *Documented findings and publication:* The resulting artifact and its evaluation must be carefully documented. Publishing the outcome, for example, in an article is an essential aspect of this.

A distinguishing characteristic of writing in technology science is the focus on describing the new artifact. Therefore, article writing also becomes slightly different from other sciences, as explained in Chapter 13.

## 3.4 Comparison with explanation science

As pointed out several times, this book distinguishes between technology science and explanation science. Explanation science aims to understand reality as it is, while technology science extends reality with *new* inventions in the form of artifacts. Explanation science may study artifacts, namely already *existing* artifacts. A social scientist may, for example, analyze the impact of social networks on education. Furthermore, what would archeology be without artifacts?

The driving force of explanation science is the desire to understand reality, while the technology scientist is motivated by artifact needs.

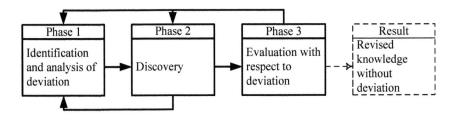

**Fig. 3.4** Overall process of explanation science.

Figure 3.4 illustrates the three main phases of explanation science. They can be summarized as follows:

- Phase 1 – *Identification and analysis of deviation:* New knowledge is required either because we lack knowledge about some observed phenomenon or because established knowledge is inconsistent with specific observations of reality. In both cases, there is a deviation between what we *know* and what we observe.
- Phase 2 – *Discovery:* The aim is to discover the true nature of the deviation and thereby adjust or extend existing theory, explanations, and understanding to fit with observations. If successful, we hypothesize that the suggested adjustments or extensions eliminate the deviation.
- Phase 3 – *Evaluation with respect to deviation:* The hypothesis is evaluated by deducing and testing predictions about observable phenomena – for example, the outcome of experiments.

To explain the three phases further, we look at two examples from the history of science. We start with the discovery of blood circulation [32], [41].

*Example 3.2 (Blood circulation).* According to established medical theory in the sixteenth century, blood was produced in the liver, transported around the body through the veins, and used up as nourishment to the tissues along the way. Moreover, the arteries transported vitality from the lungs into the body. William Harvey (1578–1657) measured the heart's pumping capacity and deduced that the human heart pumped out 260 liters per hour. This did not fit well with the blood being produced and used up without circulating, for how can the liver produce 260 liters of blood per hour? Harvey also discovered that the valves in the veins are one-way and only allow flow to the heart – that is, in direct contradiction to the theory of the time. So here, there was a large discrepancy between existing knowledge and what one could observe. A success criterion for Harvey was, therefore, to develop a theory of blood circulation that, contrary to the existing one, was compatible with observations.

William Harvey postulated the hypothesis that the blood circulates and the veins carry back what the arteries carry out. He tested the idea by tying a rope around an arm. The veins are closer to the surface than the arteries, so by loosening the rope a little, the blood could flow down through the arm but not up via the veins, which caused the veins below the rope to swell.

From a research methodological perspective, we may argue that Harvey tested his hypothesis along the following lines.

- *Existing knowledge:* – A rope around a human arm must be tightened substantially harder to squeeze the arteries than the veins because the arteries lie deeper under the skin than the veins.
- *Prediction:* – If you tie a rope around a person's arm and loosen the rope slightly, the veins below the rope will swell.

Existing knowledge in which we have great faith is postulated as a premise. The prediction forecasts the outcome of a concrete test. Harvey's test represents an attempt to falsify the hypothesis. The premise is an assumption whose truth is a prerequisite for the test. The predicted outcome of the test is a logical consequence of the hypothesis and the premise. This means that if the prediction fails, then either the hypothesis or the premise is false. If the prediction fails and we have great faith in the premise, we can conclude that the hypothesis is probably false.

Another famous example from explanation research is the discovery of Neptune [32].

*Example 3.3 (The discovery of Neptune).* In 1781, the German-English astronomer William Herschel (1738–1822) discovered the planet Uranus. Pretty soon, there turned out to be certain problems with the orbit of Uranus: It was not as should be according to Newton's laws. Discovering the cause of this discrepancy between what could be observed and existing theory was a challenge. In 1824, the German astronomer and mathematician Friedrich Wilhelm Bessel (1784–1846) hypothesized that Uranus's irregularities were due to the gravitational field of an unknown planet.

In a note presented to the French Academy of Sciences on August 31, 1846, the mathematician Urbain Le Verrier, who we know from Example 2.5, postulated the orbital parameters of the new planet and its angular diameter. Le Verrier's hypothesis was far easier to evaluate, not least because the planet should, even in a moderate-sized telescope, emerge as a crescent and not just as a star-like point. On September 23, 1846, Johann Gottfried Galle (1812–1910) at the Berlin Observatory pointed the telescope at the predicted position: Right ascension 22 hours and 46 minutes and declination –13 degrees 24 minutes. Less than 1 hour later, Galle described a star of size class 8 at a position near the predicted position that his assistant was unable to find on the star map. The next night they observed that "the new star" had moved on, indicating that it was a planet.

Although Le Verrier was wrong about Vulcan, he was right in the case of Neptune. Probably the fame of the latter discovery led to the great interest in Le Verrier's hypothesis about the existence of Vulcan.

So far, we have emphasized the *differences* between explanation and technology science from a process point of view. However, Figures 3.1 and 3.4 suggest that they also have a lot in common and follow a related pattern. Table 3.1 highlights this pattern:

**Table 3.1** Explanation science versus technology science

|  | **Explanation science** | **Technology science** |
|---|---|---|
| **Problem** | Deviation between theory and observed reality | Deviation between available artifacts and the needs |
| **The solution aimed for** | New knowledge about reality | New artifact |
| **The solution should be compared to** | Relevant part of reality | Artifact needs |
| **Overall hypothesis** | New knowledge corresponds to observed reality | New artifact satisfies the artifact needs |

In explanation science, the challenge is to explain the lack of correspondence between existing knowledge and observed reality, while in technology science, there are unfulfilled artifact needs. The solution sought in the former is extended or revised knowledge about reality, while it is a new artifact in the latter. Evaluation is required to determine to what extent the problem has been solved:

- In explanation science, we compare the proposed new knowledge with observations of the relevant part of reality. The overall hypothesis is that the proposed new knowledge explains or reflects the observations.

- In technology science, we compare the proposed new artifact with the artifact needs. The overall hypothesis is that the new artifact satisfies the artifact needs.

In both variants, we can express the general hypothesis as *B* solves problem *A*. *A* is either insufficient knowledge or unsatisfied artifact needs. *B* is the hypothesized new knowledge or artifact representing a solution to the problem.

## 3.5 Comparison with action research

Social structures such as companies, institutions, work processes, and procedures are also artifacts. Humans design them to meet human needs. Social structures are generally very demanding to evaluate. It may be difficult, and even practically impossible, to conduct repeatable studies due to interference that is only partly understood and hard to control.

Action research is a methodology designed to offer support in such difficult situations.[2] A bit simplified, action research consists of a two-step process:

1. Analysis of a social situation carried out jointly by researchers and involved parties.
2. Change experiments and effect analysis conducted jointly by researchers and involved parties.

The action researcher is usually not the problem owner. Hence, there is both a social structure (business, organization, or community) to improve, in the following referred to as the research object, and a client to relate to. The client is not necessarily a representative of the research object, but we assume this is the case to keep things simple in the following.

The first task is to arrive at a common understanding of the problem to be solved. Then the actions that should be taken to remedy the situation must be identified. The overall hypothesis is that executing these actions will reduce or eliminate the problem. The evaluation consists in performing the actions and assessing their effects. The researcher and other involved parties (representatives of the research object) cooperate in this as a team. Therefore, the action researcher is an actor interfering with the research object, not an outside observer. It is often a difficult task to balance.

The actions may involve a change of roles, responsibilities, and tasks of individuals or require them to develop new skills. Other possibilities are revising the

---

[2] Action research originated in the mid-twentieth century in social psychology [51], [84] and was later used more generally in social science and medicine. There are many variants of action research. Examples are action learning, participatory observation, and clinical fieldwork. Although action research refers to the term research, some of what is referred to as action research is more in the direction of consulting. The action research literature is not precise in terms of its basic terminology. The term action research is used both to denote a whole class of social science action methods and to characterize a subclass of these [4].

organization's structure, systems, processes, or procedures. A thorough assessment is necessary both before and after the execution of the action.

Observing and analyzing the effects of the actions systematically and critically is essential. Did the actions have the desired impact in isolation? If yes, was the effect on the diagnosed problem as expected? Was the problem solved or reduced due to the actions or other causes? Another type of learning applies to the particular action research framework used. To what extent was the framework helpful for this type of problem, and in what way should it be adjusted based on our experience?

If the actions solve the problem, the client is probably content. If not, a reassessment with a new action plan could be needed. Maybe we should also reformulate the problem. Action research is thus an iterative process.

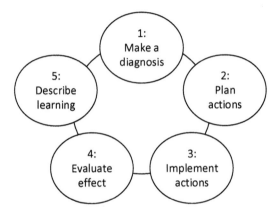

**Fig. 3.5** Five phases of action research.

Figure 3.5 presents the five phases of action research according to [81]:

1. *Make diagnosis:* Identify problems and the underlying causes of the desire to improve or change. The result is a description of problem areas and the need for improvement.
2. *Plan actions:* The actions to solve or remedy the problem are planned.
3. *Implement actions:* The actions are performed according to plan.
4. *Evaluate effect:* The effects are analyzed and evaluated regardless of the outcome.
5. *Describe learning:* The new knowledge we have learned is documented.

The researcher and the involved parties work as a team throughout the entire cycle.

At first glance, action research may seem to follow a pattern different from the processes of technology and explanation science. But if we inspect the five phases more carefully, we may illustrate action research as in Figure 3.6. Making a diagnosis is equivalent to characterizing the problem; planning actions is equivalent to establishing the hypothesis, and performing actions and evaluating the effect corresponds to testing the hypothesis. Describing learning equals writing.

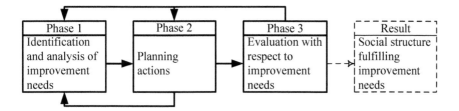

**Fig. 3.6** Overall process of action research.

Table 3.2 links the process of action research in Figure 3.6 to that of technology science in Figure 3.1. The starting point in both cases is a set of needs. The solution aimed for is an improved social structure that meets the needs for change and a new artifact that satisfies the artifact needs, respectively.

A social structure can also be understood as a human-made object and thus, in general, an artifact. In both variants, the overall hypothesis is *B* solves the problem *A*, where *B* is a new artifact.

**Table 3.2** Action research versus technology science

|  | **Action research** | **Technology science** |
|---|---|---|
| **Problem** | Deviation between current and desired social structure | Deviation between available artifacts and the needs |
| **The solution sought** | Revised social structure | New artifact |
| **The solution should be compared to** | Improvement needs | Artifact needs |
| **Overall hypothesis** | Revised social structure satisfies the improvement needs | New artifact satisfies the artifact needs |

Thus, to the extent we are willing to accept a social structure as an artifact, we may view the process of action research as a particular case of the process of technology science. That is the case where the artifact is a social structure, and the researcher plays a role in the object of investigation.

Our example of action research is taken from anthropology, namely a study of the Meskwaki people, a native American people from the Midwest in the United States [60].

*Example 3.4 (The Meskwaki people).* A variant of action research is known as action anthropology. Action anthropology arose from a basic research project initiated by Sol Tax (1907–1995) at the University of Chicago in 1948. The goal was to study the Meskwaki people (also known as the Fox people), but as the research progressed, the researchers became interested in the problems of the Meskwaki people at the time. From this, a desire arose among the scientists to help the Meskwaki people to deal with some of these problems.

The Meskwaki people have traditionally perceived themselves as unsuccessful in dealing with people of European ancestry. In the value system of Americans of European origin, the ideal is to realize oneself through hard work and achievement. On the other hand, the Meskwaki people were content to accept themselves as they were. They were motivated by external factors and not internal moral principles. They sought social acceptance and feared condemnation. This attitude had traditionally been interpreted as laziness by white Americans. Consequently, the Meskwaki people felt guilty and perceived themselves as unsuccessful.

Two anthropological attempts were made to remedy this situation. The first was about enlightenment aimed at the white population, while the other focused on encouraging the Meskwaki people to help themselves. In the first case, the hypothesis was that enlightenment aimed at white Americans would change their attitude towards the Meskwaki people, reducing the latter's guilt.

In the second case, they would assist selected members of the Meskwaki people with artistic talent in establishing a successful craft business. They hypothesized this would increase their confidence because it would prove that they could operate effectively within the US economy and increase their revenues substantially.

## 3.6 Comparison with technology development

In the technology domain, research and development are closely intertwined. Many, perhaps most, research projects in technology science will include aspects or components of technology development – aspects that cannot be classified as research. On the other hand, many projects are about development without any research. Most industrial projects belong to this category.

We may perform the following test to determine whether we are doing technology science or pure technology development. It consists of answering three questions:

1. Does the artifact (or an aspect of the artifact) represent new knowledge?
2. Does the new knowledge interest others?
3. Is the new knowledge documented in such a way that others can check it?

If we answer yes to all three questions, then we are doing technology science. In the opposite case, we are pursuing pure technology development.

*Example 3.5 (Acquisition system).* A large company needs to renew its IT system for procurement. It turns out that no commercially available system satisfies their needs. Nor do other companies of similar size or kind (as far as they know) offer a system that fulfills their expectations. Therefore, they decide to build significant parts of the procurement system from scratch. The company's software developers work with relevant stakeholders throughout the company to specify the new system. The system is ready for use after many rounds of programming, testing, and requirements tuning. It turns out that most users are satisfied with the result because their working routines have been simplified, and they now have a better overview. Furthermore, due to its efficiency, they now have time for previously neglected tasks.

The resulting artifact, the new procurement system, was successful in this case. Now let us try to answer the three questions formulated above:

1. Does the new artifact (or any aspect of it) represent new knowledge? Looking at the new procurement system separately, it is conceivable that this system is the only one of its kind. But that is not necessarily crucial. The essential question is whether any part of the system (or an aspect such as its architecture) represents new knowledge and whether it can be significant to others who will create similar designs. It leads to the next question.
2. Does the new knowledge interest others? The new procurement system may not be of interest outside the company in question. On the other hand, the system or some aspect of it may be of interest if the principles can be reused in different contexts.
3. Is the new knowledge documented in such a way that others can check it? Success stories without proper documentation are only loose statements. Research requires verifiable documentation. That is, the quality of the documentation must allow others to test and check the research.

A result that satisfies the above three criteria represents research and is thus worth publishing. The point of publishing is to make the findings known to others, including other researchers and potential stakeholders. That way, the result can be debated and possibly criticized, as well as contribute to further research and development. Therefore, the dissemination process is of great importance for the research community and society. We will return to that in Chapter 12.

## 3.7 Hybrids of different types of research

As explained above, many research projects in technology science involve aspects of development. Therefore, there is a sliding scale between technology science and technology development. Furthermore, many research projects in technology science include sub-projects that belong to explanation science and the other way around. A typical example of the latter is the exploration of space. To map the distant regions of the universe, you need advanced instruments like the Hubble Telescope.

The mapping is explanation science, while the construction of the specialized devices is technology science.

Although many research projects involve combinations of different types of research, it is nevertheless vital to be able to distinguish them. One reason is the choice of research methodology. Technology science employs other research methods than, for example, explanation research. Another reason is publishing. In technology science, a research article will devote much space to describing the artifact or invention in question. In explanation science, the focus is on the improved understanding of reality. The most important reason, however, is the need for precision. When starting a new project, it pays to divide the project into its components and specify their deliverables. The project mentioned above focusing on the distant regions of the universe, for example, should be divided so that the explanation and technology parts can be conducted separately with a clearly defined interface.

A famous example of how explanation research can support technology research is Semmelweis's introduction of cleaning procedures to reduce the occurrence of puerperal fever.

*Example 3.6 (Puerperal fever).* Many research projects in medicine consist of two main components. The first is explanation oriented and is about identifying the cause of some medical problem or illness. The second is technological and aims at inventing artifacts to remove or reduce the cause or consequences of the first. The artifact can be a new medicine, procedure, a guideline for what to do or not to do, or a combination.

The research of the physician Ignaz Semmelweis (1818–1865), who worked in Austria- Hungary, can be seen as an example of this. Semmelweis was worried about the considerable proportion of newly-born mothers who died of puerperal fever at a particular maternity ward at the hospital where he worked. In this department, medical students helped the women during birth. At the same hospital, there was another maternity ward serviced by midwives, where mortality was far lower. Semmelweis hypothesized that puerperal fever was due to some substance the medical students unknowingly transferred from the autopsy room to the maternity ward. He hypothesized that handwashing with a chlorine lime solution would be sufficient to remove any infectious material that the medical students got on their hands during an autopsy. Based on this assumption, Semmelweis predicted that if the puerperal fever were due to an infectious agent from the autopsy room, the mortality rate would be reduced if the students washed their hands in advance with a chlorine lime solution. Semmelweis instructed students to wash their hands with chlorine lime solution before entering the maternity ward to test the prediction. After this, the mortality among the mothers in the ward fell from over 12% to just over 2%, comparable to the second maternity ward figure. So the prediction turned out to be correct.

Semmelweis's research has an explanation aspect because the hypothesis represents new knowledge about reality as it is. It is also technology science, as the new routine or procedure for cleaning hands is an artifact invented as part of the project. The remarkable thing about Semmelweis's research is that the artifact is relatively simple and easy to understand, contrary to today's medical research. Another interesting phenomenon with Semmelweis's discovery is that the artifact has two roles. First, it gives Semmelweis a solution to his practical problem. Second, it is used to test the explanatory hypothesis.

## 3.8 The role of machine learning

Machine learning (or ML for short) is widely used for all kinds of scientific work. Some [26], [30], [7] argue that ML is also changing the way we do research and the scientific method as such. It therefore makes sense to discuss the relationship between ML and what has been presented in this chapter.

ML augments systems with capabilities that seemingly mirror those of human intelligence. These capabilities may improve the effectiveness and automate tasks that previously required human intervention. ML [66] is mainly good at low-level pattern recognition when analyzing data sets too large to make human processing practical.

To build ML, we need data. The data is split into training data and test data. Roughly speaking, the training data is used to teach the software to become "intelligent," and the test data is used to check whether the software became "intelligent." The training may take place offline, online, or both offline and online. Hence, the data is not always collected in advance.

The final result of the training process is known as the ML model. The ML model is a software program that analyzes input and produces output in an "intelligent" manner. The input could be, for example, an X-ray image, and the result could be a probability that the patient in question has cancer. Typically, users send input data to the ML model and receive output without understanding the internal workings of the algorithm.

It is essential to distinguish between the researcher employing existing ML technology as a research method and the researcher trying to improve ML as a methodology. In the following, we are concerned with the former. We also restrict ourselves to the invention phase, but ML can be used for other purposes such as analyzing needs and evaluating hypotheses.

The ML model is an artifact like any other software program. However, its engineering is different because it represents the output of another software program that has performed the learning process. Human engineering of ML means engineering the learning process. The learning process is also an artifact.

The researcher designs the learning process for the problem at hand. It typically involves selecting the ML approach, extracting data sets, defining reward functions, etc. The suitability of the design chosen for the problem of relevance is an implicit working hypothesis. If the design is good, executing the learning process results in an ML model that solves the problem. A bit simplified, we can understand "the ML model solves the problem in question" as a hypothesis about the generated software.

This hypothesis may be wrong, like any other hypothesis. The data may, for example, have been insufficient, poorly selected, or manipulated, and the software may suffer from all kinds of hiccups. Hence, a hypothesis generated by ML requires the same kind of careful evaluation as a hypothesis invented by a human before it can be considered trustworthy.

We use Einstein's mass-energy law as a basis for comparison. We do not know the details of Einstein's thoughts, but at some point, he must have had a working hypothesis along the following lines:

> A system's energy $E$ and mass $m$ and the speed of light in vacuum $c$ can be related by some sort of mathematical equation to reflect reality.

After much thinking and many thought experiments, Einstein formulated the following hypothesis:

$$E = mc^2$$

Now, let us consider a researcher using ML as a research method. Our source of inspiration is the use of ML to make the chemical process of soap production more efficient and, in addition, reduce pollution [29] by predicting a neutralization number. A working hypothesis could be:

> For a chemical sulphonation process, it is possible to come up with an ML model that predicts the neutralization number with sufficient accuracy.

In this hypothesis, the ML model (to be made) plays the same role as the equation (to be formulated) in the working hypothesis for the mass-energy law. Assume, for simplicity, `DesignDoc` describes the final design of the learning process (data included). It leads to the following hypothesis:

> For a chemical sulphonation process, the ML model generated as specified in `DesignDoc` predicts the neutralization number with sufficient accuracy.

Again we see that the ML model represents the equation in Einstein's law. Contrary to the ML model, the equation is exact, but except for that, they play the same role. Einstein came up with the equation himself, while the ML model was generated automatically by the designed learning process. Future tests may falsify both hypotheses.

To summarize, ML assists the researcher by automatically generating the ML model based on the researcher's design but does not change the scientific process. The researcher still needs to characterize the problem to be solved. Furthermore, the researcher must define the research method in the form of a working hypothesis and conduct careful evaluations to check whether the ML model solves the problem.

# Chapter 4
# Problem Analysis

To solve a problem, we must understand its true nature. Problem analysis is a means to clarify the issue and devise a plan for how we should approach it. In technology science, this means identifying and characterizing the artifact needs and deciding on a strategy or way forward to fulfill them. This and the following chapter will closely examine what this may involve.

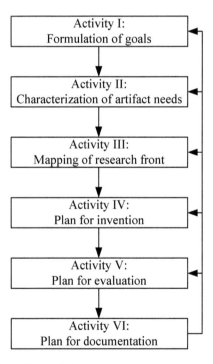

**Fig. 4.1** The activities of problem analysis.

Figure 4.1 gives an overview of the various elements of a problem analysis. The figure distinguishes between six activities. These six activities represent a decomposition of Phase 1 of Figure 3.1, the needs identification phase. Activity I is concerned with clarifying the research goals. What effects do we hope to achieve in the long term (beyond the project lifetime), and what concrete results (new artifacts included) do we aim to deliver within the project lifetime? Activity II aims to identify and provide a precise characterization of the artifact needs. If we are to succeed, we depend on a good understanding of what these are, and this understanding must be documented appropriately.

Activity III focuses on mapping the research front – establishing the state of the art. This includes obtaining an overview of existing technology, knowledge of relevance, and how it can be used or utilized.

The remaining three activities address planning. Activity IV concerns the invention phase. In other words, how we intend to create, establish, or design new technology to meet the artifact needs. Activity V is dedicated to the evaluation phase – how to evaluate the new technology. That is, within given financial and other project constraints, how do we plan to assess if or to what degree the artifact needs have been fulfilled. Our research must be adequately documented to allow others to check and test it. This also requires planning, and this is the task of Activity VI.

Figure 4.1 presents the problem analysis as a sequence of six activities. In practice, we will often go back and forth between activities. For example, it is not uncommon to go back and adjust the research goals after characterizing the artifact needs. The figure illustrates this by the feedback loops. This chapter addresses Activities I-III, while Chapter 5 is devoted to Activities IV-VI.

## 4.1 Formulation of goals

A goal is always linked to a specific stakeholder. The stakeholder can be an individual or, for example, an organization.

**Definition 4.1** A *stakeholder* is one participating in or having interests in some enterprise.

Different stakeholders may have various goals. Often, technology science is funded jointly by companies or businesses and the public sector, such as a public company, a ministry, or a research council. A business may aim at making money, while a public company may want to strengthen the competence in a particular field or contribute to solving societal problems. The researchers themselves are also stakeholders. One may be hoping to obtain a Ph.D. based on results from the project, while another intends to contribute to the best of humanity by establishing new environmentally friendly technologies.

The terms *goals* and *objectives* are widely used and not specific to research. They are also employed interchangeably. Moreover, it is common to distinguish between outcome and impact goals.

> **Definition 4.2** An *outcome goal* describes what a project or measure should achieve and is linked to the direct results and outputs of the project. An *impact goal* describes why the project has been established, for example, helping to reach a desired future societal state (in many cases, not within the lifetime of the project).

In some contexts, objectives are interpreted as outcome goals and goals as impact goals. In this book, however, only the term goal is used, and we distinguish between outcome and impact goals.

*Example 4.1 (Outcome and impact goals for a space project).* A research project to position a satellite to orbit Jupiter may have the following impact goals:

- Improve understanding of the origins of the solar system.
- Increase knowledge about Jupiter and its moons.

Possible outcome goals, on the other hand, are:

- Position the satellite in orbit.
- Succeed in making all instruments work according to their specification.
- Implement the project within available budget constraints.

Whether the outcome goals have been reached can be checked at project termination. This may not be the case for the impact goals. The satellite in the above example can gather data for many years. Hence, the impact goals depend on data collection and subsequent analysis long after project termination.

## 4.2 Characterization of artifact needs

Activity II concerns the identification and characterization of artifact needs. The aim is to establish a high-level understanding. We are not supposed to deliver a requirement specification. However, the latter may be included as a task in the plan for the invention phase (as described in Section 5.1.2).

As Figure 4.2 illustrates, the needs characterization can be decomposed into three sub-activities: identification, analysis, and documentation of artifact needs. These sub-activities are primarily parallel, but in the following are addressed separately.

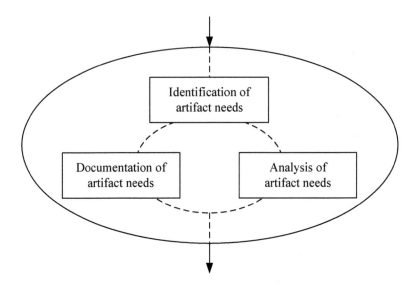

**Fig. 4.2** Characterization of artifact needs.

## 4.2.1 Identification of artifact needs

Different stakeholders have different needs. When identifying needs, the first step is to identify the relevant stakeholders.

*Example 4.2 (Stakeholders for a time machine).* Traveling in time involves moving to another reality, in either the past or the future. The latter we know is possible. If we are exposed to increased gravity or speed, time will slow relative to others who are not exposed to this. When Sergei K. Krikalev (b. 1958) ended his career as a cosmonaut in 2005, he had spent a total of two years, ten weeks, two days, and 22 hours in space [25]. Due to the time effects related to gravity and velocity, Krikalev was then 0.02 seconds younger than he would have been if he had lived an everyday life on Earth. In other words, he had traveled 0.02 seconds into the future. The extent to which it is possible to travel back in time is disputed [35].

Time machines are well known in science fiction. The characters travel both backward and forward in time. Suppose that such journeys are physically possible and that we have been commissioned to build a time machine.

Who are the stakeholders? Well, the time traveler is a stakeholder. For the time traveler, personal safety is most likely a concern. Second, to the extent the time machine is operated from the reality constituting the journey's starting point, we must pay attention to the needs of the operators. Ease of use and minimal health risks are probably crucial for the operators. Third, since changes to the past may affect the future, humanity, represented by some authority, is also a stakeholder. That time travelers can only observe and not modify the past is a natural expectation. Fourth, an investor might hope to make money on the technology, for example, by offering time travel for a fee. The investor may want to keep the costs low.

A time machine would be spectacular, but everyday technology science is much more down-to-earth. The artifact to be created is often intended to replace one or more already existing artifacts, for example, processes or procedures already in use. If so, studying the workings and weaknesses of these existing solutions is a widely used and highly recommendable strategy to identify needs.

Digitalization is thus a good example. Digitalization projects aim to improve, assist, or replace existing approaches using computerized technologies. In the following, we look closely at a case from the offshore industry.

*Example 4.3 (Digitalization in offshore).* During maintenance projects on offshore installations, it is common to accommodate personnel in floating hotels – so-called flotels. The personnel use a mobile bridge to go back and forth between the flotel and the installation. When the weather is terrible or exceptional situations occur, the bridge is lifted to avoid material damage or human exposure to danger. Whether or not the bridge should be raised can be challenging to decide, as this depends on many factors. In addition, there are also financial consequences associated with this.

Human operators decide whether to lift the bridge and so it will continue for years to come. Nevertheless, the operator can benefit from computerized decision support. Assume we have been commissioned to create such a decision support tool. Stakeholders, in this case, are users of the bridge and operators, but also representatives of the flotel and installation. Additionally, a software vendor may be aiming to design a tool that can be used broadly across multiple sectors to maximize revenue. To identify and understand the artifact needs, a study tour of an installation with an ongoing maintenance process may be helpful. Interviews with operators and personnel are also relevant to identify which factors are crucial when deciding to lift the bridge.

In many projects, a study tour, as described above, is too time-consuming for the problem analysis. It can then instead be integrated into the plan for the invention phase.

### 4.2.2 Analysis of artifact needs

As the needs characterization proceeds, it can be helpful, or even necessary, to tidy, structure, and clarify the relationship between the different expectations. It is not uncommon to discover that the captured needs address not just one artifact, but several. On other occasions, if we aim for several artifacts, the problem may be that their needs are mixed. It is advisable to keep the different artifacts apart and identify their needs individually. It can also be helpful to break down large artifacts into sub-artifacts and address them individually.

Needs may contradict each other, partly or wholly. What is helpful for one stake-holder may be problematic for another. Needs can also be incompatible because stakeholders expect too much. Security needs can, for example, make it challenging to fulfill requirements for usability or privacy.

Suppose the research project involves components of both technology and expla-nation science, which is the case more often than not. In that case, it is advisable to treat these as separate sub-projects without losing the overall picture. In Figure 4.3, the artifact needs of technology science from Figure 4.2 have been replaced by the knowledge needs of explanation science.

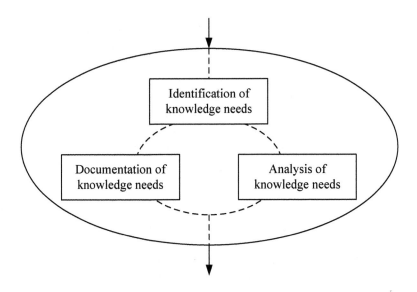

**Fig. 4.3** Characterization of knowledge needs.

The breakdown of a project into sub-projects or tasks, as indicated above, simplifies the needs capture as well as the research planning. It can, however, be challenging to know when to stop – in other words, to decide on the level of granularity. Ex-perience is essential here, and it is difficult to give general guidelines. If you are inexperienced, the following rule of thumb may help: Do not decompose into com-ponents so small that their results cannot be presented independently in stand-alone reports.

### 4.2.3 Documentation of artifact needs

When working with others, written documentation is always necessary. Even in one-researcher projects, although the artifact needs seem clear, writing them up on paper or electronically is advisable. First, it is easy to forget what we once thought we understood. Second, when expressed as something more than a thought, we may discover issues or problems we were unaware of.

Artifact needs may be documented in many ways. As long as they are described correctly, sufficiently precisely, and in a manner that is easy to understand for the stakeholders involved, the exact form is less important. To achieve these qualities may, however, be challenging.

Some express artifact needs in the form of success criteria.

**Definition 4.3** A *success criterion* characterizes a test or condition that must be fulfilled to succeed with a given enterprise.

In technology science, needs may, for example, be expressed as expectations of the new artifact that the research should result in. We may think of such success criteria as high-level but reasonably precise requirements. As a rule of thumb, they should be presentable in the introduction to a technical report documenting the research results. Hence, they should not depend on much technical background.

*Example 4.4 (Success criteria for a time machine).* An obvious success criterion for the time machine from Example 4.2 is that it works:

> The time machine can transport a human being to some other time in the past or future and back again.

If we develop new technology, we are usually concerned about its safety. This can be challenging for something as groundbreaking as a time machine, and therefore, the following success criterion seems essential:

> The time machine is safe in the sense that it does not harm the time traveler, its operators, or the surrounding environment.

Most of us would be very concerned if time travel could potentially affect the reality we live in by changing the history of the past. We may think of horror scenarios where the time traveler kills one of our ancestors or helps a party win an election or war that this party lost. It, therefore, seems natural to require:

> Using the time machine will not affect the reality or time period being visited.

Although not much emphasized in science fiction literature, there are privacy challenges associated with time travel. What if someone travels back to spy? For many of us, there is perhaps not much juicy stuff to find, but we still value our privacy. Hence, the following makes sense as a success criterion:

> The time machine is configured so that time travel does not violate current privacy principles.

There are many more expectations and requirements for a time machine than those expressed in the four criteria above. Still at an overall level, as part of the problem analysis, this may be sufficient. We should, however, plan a requirements capture at a later stage.

Success criteria are just one of several means to capture needs. If you struggle to arrive at a good description, you may attempt the following procedure.

**Procedure 4.1** *Characterization of artifact needs:*

1. Formulate a single sentence that describes at a general level what the new artifact should be good at.
2. This sentence will typically contain terms and phrases that are vague and open to interpretation. There may be concepts that are not sufficiently precise or used inconsistently. There may also be references to scales or values that are ambiguous. The lack of limit values or parameters is a possible weakness. For each such issue of concern, add one sentence of the form below to increase precision:

    By [...] is meant [...].

*Example 4.5 (Procedure applied to decision support tool).* We use the support tool case from Example 4.3 to demonstrate the procedure. The tool to be invented can be described in one sentence as follows:

The tool should, given access to relevant data, provide timely, helpful, and user-friendly advice on whether to lift the bridge or not.

The sentence gives a pretty good description but uses concepts whose interpretation is open, which is quite common when something nontrivial is expressed concisely. For example, what is meant by timely advice? We know that calculation takes time. Since the recipient of the advice is a human being, and not another computer program or machine, the following is likely to be sufficient:

By *timely advice* is meant *the operator does not feel that latency delays decision-making.*

The quality of the output of a computer program depends on the quality of the data it is fed. In other words:

By *access to relevant data* is meant *the tool is provided with the information on wave height, current, and wind speed needed to provide adequate advice.*

The exact nature and format of the input data must be specified further at some point, but this is not necessarily an issue for the problem analysis. For the advice to be practical, the output must, however, be sufficiently correct. In particular, the guidance should not lead the operators to make mistakes they would otherwise avoid.

By *useful advice* is meant *the tool gives valid recommendations that potentially lead to better operator decisions.*

By *user-friendly advice* is meant *the tool is easy to use, and the recommendations are unlikely to confuse or mislead the operator.*

The above clarifications also contain terms or formulations whose meanings are ambiguous. For example, what does "unlikely to confuse the operator" mean? These can be made more precise with new clarifications to the extent desirable. But again, it is essential to remember that the aim during problem analysis is a high-level description.

Generally, if multiple artifacts are to be invented, the artifact needs to be described separately for each of them. Understanding the needs increases the chances of a successful outcome.

A proper description of needs will also simplify reporting and article writing. If we struggle with the discussion when writing a scientific article, it is often because the artifact needs are poorly characterized or inaccurate. We will return to this in Chapter 13.

## 4.3 Mapping of research front

When we approach a new problem, it is vital to identify the aspects of the problem that others have already solved, what parts of the problem we will focus on ourselves, and what aspects of the problem we leave to others or future research. This requires a good overview of the research front, including the already available technology. If we lack this knowledge, it must be acquired. Considering time restrictions and a limited budget, it is crucial to read efficiently. We recommend reading in parallel with the other activities constituting the problem analysis. It is almost as bad to read too much as too little or irrelevant material.

During the problem analysis, we aim primarily at an overall overview of the research front. In technology science, this typically involves evaluating existing technology at a relatively high level of abstraction. This evaluation may conclude that a detailed review of certain technologies should be included in the plans for the invention phase.

There are many guidelines and procedures for conducting literature studies and evaluations. For example, [31] overviews 14 study types with associated methods. As in the case of [96], the focus is on health-related research, but the methodology is generic. The same applies to [50] and [44].

# Chapter 5
# Planning

Chapter 4 decomposed the problem analysis into six activities (see Figure 4.1). Moreover, we addressed the first three of these: *formulation of goals, characterization of artifact needs,* and *mapping the research front* in further detail. This chapter does the same for the *invention, evaluation,* and *documentation plans.*

Technology science is about creating, establishing, or inventing artifacts to fulfill human needs. How to proceed depends on the people and institutions involved, the kind of technology in question, time frame, budget, and so on.

The appropriate level of planning detail varies accordingly. A project to invent a vessel for manned interplanetary space travel requires an entirely different level of planning than a project to solve a mathematical equation algorithmically. A spacecraft depends on large-scale collaboration, while the algorithm, depending on the kind of equation, can be worked out by a single researcher.

Many technology science projects include sub-projects belonging to explanation science. For example, although the artifact needs are captured by the problem analysis (as described in Chapter 4), a sub-project to identify detailed requirements may be necessary. A sub-project to provide additional knowledge on the properties of materials and substances to be used in the new artifact may also be required. Such sub-projects should be described and planned separately, but not without keeping the entire project in mind.

We distinguish between three different plans. First, we may need a plan for the invention phase. That is, how to go about establishing the new technology. Second, when the technology matures, it must be evaluated. Again, planning is required. Third, the artifact, the evaluation layout, and the outcome must be documented to facilitate testability and repeatability. This also requires planning.

A technology science project may also need other plans. For example, plans for dissemination and knowledge transfer beyond purely scientific publishing, or commercialization and innovation. These are all important but outside the scope of this book.

© The Author(s), under exclusive license to Springer Nature Switzerland AG 2023
K. Stølen, *Technology Research Explained*, https://doi.org/10.1007/978-3-031-25817-6_5

## 5.1 Plan for invention

The need for planning depends, as already explained, on the project in question. This is especially true for the invention phase. A plan should support idea generation and creativity but, at the same time, give sufficient room and freedom. There is a broad spectrum of tutorials and books presenting or offering methods and techniques for creativity. Most of these are specific neither for invention nor research but can still be used by a technology scientist. In the following, we limit ourselves to the most essential, and especially what the author of this book has benefited from in his research.

### 5.1.1 Idea generation

We need good ideas to formulate interesting hypotheses, which in turn require the ability to generate ideas. Humans all have different talents, and no one is equally good at everything. Some run 100 meters significantly faster than others. Nevertheless, although most of us will lose against a top sprinter, we will run faster with the proper training and a better diet. Analogously for idea generation, some are initially more creative than others, but most may increase their creativity by practicing and using suitable methods and procedures. This requires effort and willingness to try out and test new things.

Facing a scientific problem is a bit like surveying foggy terrain. The terrain hidden by the fog is the new knowledge or the new artifact that we are trying to envision. Sometimes we may glimpse something that disappears before we understand what we saw. Other times we see a little more or seem to recognize something we have seen before but are unsure what we see or whether we are just fooling ourselves.

An idea often appears as a flash in the fog. A famous example in this respect is how Wilbur Wright came up with the idea of controlling an airplane by turning the framework supporting the wing pair [10] from playing with the box of a bike cable while talking to the customer he had just sold it to.

In a research situation, such glimpses appear suddenly in the most unexpected situations. A single glimpse is not necessarily valuable. Most are not helpful in themselves, but by addressing such glimpses, for example, by trying to document or illustrate them, we can trigger new glimpses, which may stimulate further idea generation.

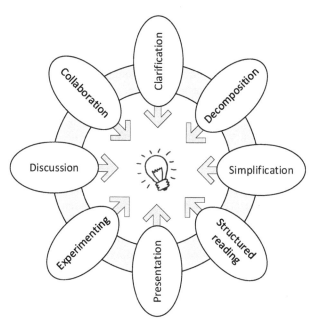

**Fig. 5.1** Idea creation.

## 5.1.2 Basic techniques for idea generation

Getting good ideas cannot be planned, but we can plan activities that facilitate idea generation. This book distinguishes between eight basic techniques or directions for idea generation (see Figure 5.1). Each of these is described in further detail below. Although presented separately in the following, they will often be combined in practice.

### Idea generation by clarification

What if we have no ideas at all? In other words, what can we do if we face a problem that we should solve and feel completely blank? Most of us are familiar with the situation. In my experience, the best way forward is almost always to try to specify the problem more precisely. A problem may appear difficult to solve because it is poorly formulated or poorly understood. Try to write it up in your own words instead of keeping everything in your head or only rereading what others have prepared. Writing things up, possibly supported by drawings, helps clarify your understanding.

Writing it up in your own words can also be fruitful when developing or refining an existing idea. It may allow weaknesses or new alternatives to appear. Moreover, ideas tend to multiply and give rise to new ideas due to the writing process.

The ideas may not appear immediately, though many do. They may pop up in two weeks or when we brush our teeth the morning after. If making the problem more precise does not help, we may turn the situation on its head and instead try to argue that it cannot be solved within the given context and available resources. In some cases, such an argument may be found, but usually, there is a path forward, and new ideas may emerge from trying to argue the opposite.

Clarifying the problem to be solved in a one-researcher project may be difficult. If the project also involves others, it is, however, a more significant challenge. In technology science, the needs to be fulfilled are an essential ingredient. Chapter 4 highlights the importance of capturing and understanding the needs; otherwise, we do not know what we aim for. Therefore, one possibility is to conduct a needs analysis to arrive at a more detailed understanding.

Performing a gap analysis is another option. This describes the discrepancy between what we have or what is possible with existing technology and what we aim for. Gap analysis is proper when the needs are partly met by pre-existing technology or when we already have a system, process, or technology that should be improved to fulfill new expectations.

The use of advanced instruments and software for data analysis can be very effective. Machine learning, as discussed in Section 3.8, is an example.

## Idea generation by decomposition

*Divide et impera* (divide and rule) is a principle of cynical leadership; it is also a problem-solving strategy. Problems can be challenging to solve because they are composed. A problem may be an aggregate of several lesser issues more easily solved individually. A problem may also be layered into sub-problems so that it is solvable only if we address the sub-problems one by one in a specific order.

Many technology science projects can be simplified by separating out elements that belong to explanation science. If we aim to invent new technology to help elderly people recover from femoral fractures, understanding the context in which it will be used and the physical constraints of the user group is essential. It may require a separate study that can be conducted as a sub-project on its own.

The new technology may consist of hardware and software that can be designed separately in two sub-projects. In this way, we can continue. If we decompose the right way, we end up with a neat and well-structured overall project, providing a fertile basis for idea generation. A project plan may consider this by splitting the invention phase into several steps or sub-projects, each focusing on a specific task or sub-problem.

## Idea generation by simplification

Ideas that prove fruitful when addressing a special case may pave the way for a general solution. To solve a challenging problem, we may therefore start by tackling a simpler version. We may, for example, add assumptions, disregard certain aspects, or impose restrictions.

Suppose we are to invent a cross-country ski offering good grip without harming the glide. Existing commercial solutions based on "fish scale" or "skin" detract from the glide. Finding the right balance between grip and glide depends on the type of snow and the temperature.

Getting grip is easy at very cold or very mild temperatures but challenging for temperatures around zero degrees Celsius. Fresh snow often provides a good grip, while old "sugar-like" snow is problematic. To simplify the task, we may first limit ourselves to a specific temperature range or type of snow. If we successfully address the particular case, we can after that examine to what extent the solution works under other conditions or can be generalized. Conversely, if we cannot solve the particular case, we have a clear indication that the overall task is too difficult.

During the planning phase, we may exploit "idea generation by simplification" by dividing the invention phase into several steps where each new step adds complexity until the general case is reached.

## Idea generation by structured reading

A problem is often related to other problems whose solutions are known or available in the literature. A possible idea generation strategy is to carefully look into these and try to characterize why our issue cannot be solved correspondingly.

Evaluating existing technology is a source of inspiration and creativity. As described in Section 4.3, Activity III involves a mapping of relevant technology. However, this mapping is at a high level of abstraction and embraces a lot. An evaluation during the invention phase is typically specialized and focuses only on the most promising technologies or technology solutions.

The level of formality of the evaluation process and the required level of planning detail are context-dependent. A structured evaluation of existing solutions based on specified criteria makes it easier to argue later why we selected or adapted one of these or found it necessary to come up with a completely new approach. However, such evaluations are resource-demanding.

A problem usually consists of many sub-problems, and conducting a structured literature study each time we get stuck is unfeasible. Reading efficiently is essential, and this is not necessarily the same as reading an article, report, or book in the way intended by its authors. On many occasions, it may be advisable to start with the main result and then read just enough of the context or the background to determine whether this is suitable for our work or not.

The problem must guide the reading. Although reading is essential, researchers often read too much or wrongly. In most subject areas, there are enormous amounts of literature, and unless we select our readings carefully, we risk reading a lot that does not move us forward and not what we should have read.

## Idea generation by a presentation

Another approach to idea generation is an oral presentation, for example, to colleagues or other appropriate listeners. The presentation may describe our understanding of the problem we are trying to solve and perhaps some ideas for potential solutions or paths forward.

One outcome of the presentation is input from those we are presenting to. Equally important, by preparing and giving the presentation, we may see our work in a new light or from a different perspective and thereby be able to move on regardless of the kind of response we get from the audience.

Immediately after a presentation, the brain usually works in high gear. It is a very creative moment and, therefore, essential to exploit. There may be aspects in the presentation itself, new associations, or phenomena we now suddenly see from a completely different perspective that allows us to find new angles of attack. There may also be questions we receive or remarks made along the way that trigger new thoughts. Write down thoughts and ideas, or illustrate them graphically.

Presentations can be organized at short notice, but for larger projects, they may be planned activities during the invention phase and later stages.

## Idea generation by experimentation

Experimentation as a means for idea generation involves trying out, testing, or simulating different alternatives to assess how things work.

In technology science, we often make prototypes to improve understanding and identify technical challenges or how potential solutions work. Such prototypes can be anything from cardboard models to advanced technical installations. Experimentation may also involve testing materials or evaluating existing technology in new environments.

Making sketches is also a form of experimentation. A sketch may, for example, outline a potential solution and is an excellent medium for idea generation. It is often helpful to operate with several alternatives. The sketches could be understood as living documents that are modified and refined until the design is completed.

Thought experiments are alternatives to conventional experiments when the latter are unfeasible due to physical, technological, ethical, or financial constraints. Thought experiments have been used as long as humans have been engaged in science. For example, Lucretius (99 BCE – 55 BCE) made use of the following thought experiment [9] to show that the universe is infinite:

"... if there is a purported boundary to the universe, we can toss a spear at it. If the spear flies through, it isn't a boundary after all; if the spear bounces back, then there must be something beyond the supposed edge of space, a cosmic wall that stopped the spear, a wall that is itself in space. Either way, there is no edge of the universe; space is infinite."

Einstein made extensive use of thought experiments. Furthermore, thought experiments were essential in the establishment of quantum mechanics. The space elevator [22] that is supposed to transport personnel from the ground to a vessel in space is an excellent example of a thought experiment that has gradually grown into something more and that will probably be realized in the future.

During the planning, it is possible to define different experimentation activities and the methodologies and tools that should be used, such as specific kinds of instruments or machine learning. It may include activities related to relevant preexisting thought experiments as well as arrangements to identify or define new thought experiments.

## Idea generation by discussion

Discussion can be so much. We can discuss with like-minded people, critics, and across disciplines. We can discuss publicly or inwardly in groups. Regardless of the form, factual and focused discussions may fertilize the idea generation process.

Criticism or disagreement is not necessarily negative. Often discussions between people with different backgrounds and beliefs are the most fruitful. Spontaneous conversations occur all the time and do not require planning. There are also formal methods and processes for conducting or steering discussions. We may plan formal sessions for idea generation with or without an instructor or tutor.

The formality may, for example, ensure that everyone gets their say or that all aspects of a problem are covered.

## Idea generation by collaboration

Discussion can be understood as a kind of collaboration and presentation to others. But ideas can also arise from more general forms of cooperation. For example, knowledge may not be utilized because researchers at a large research enterprise are unaware of the expertise offered by other researchers at the same institution.

To address this, we may plan working groups or meetings where researchers and other potential stakeholders are challenged to attack problem issues or aspects of relevance. The setup and organization of such meetings require careful planning.

How to obtain, identify, characterize, analyze, and share knowledge within a business or organization is known as knowledge management and is a giant field of study.

## 5.2 Plan for evaluation

Evaluation is required to determine whether our invention, the new artifact we have come up with, satisfies the artifact needs. Evaluation can mean so much. In this book, evaluation involves checking, testing, or assessing the correctness of hypotheses. Terms like verification and falsification can be understood as special cases. The evaluation phase is demanding and usually requires careful planning. In the following, we first give an overview and classification of evaluation methods. Then we have a closer look at some evaluation challenges of relevance for the planning phase.

It is worth noting that the methods we refer to as evaluation methods are more generally applicable than our term indicates. The same methods play other roles in many contexts and are often referred to simply as research methods.

### 5.2.1 Classification of evaluation methods

Joseph McGrath (1927–2007) [56] highlighted that all evaluation methods have built-in or inherent weaknesses. We want an evaluation that maximizes respectively

- generality,
- precision, and
- realism.

But generality, precision, and realism cannot be maximized simultaneously.

For example, concerning the categories of evaluation methods defined below and presented in Figure 5.2, logic supports generality since logical principles are universally valid. Logic is, on the other hand, non-empirical and therefore scores poorly on realism. The same goes for mathematics. A survey is suitable for gathering general knowledge but goes into little depth and is, therefore, less good at realism. Field studies and field experiments are realistic but not very precise because there are potentially many uncontrollable factors that affect the outcome. Laboratory experiments allow a high degree of precision but are based on simplifications and artificial environments and are therefore weak on realism. The same applies to a large extent to prototyping and experimental simulation. To correct the weaknesses of the various methods, they must be combined, so they cover each other's weaknesses.

Definition 5.1 provides a brief characterization of each category.

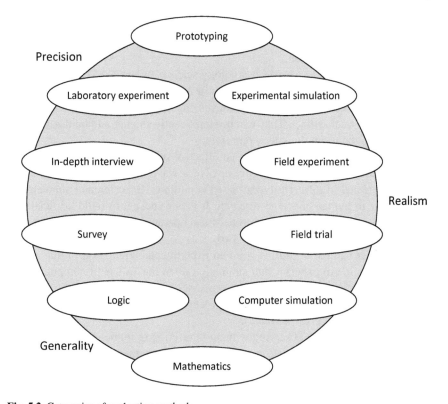

**Fig. 5.2** Categories of evaluation methods.

**Definition 5.1**

- *prototyping* – involves building a model of an artifact;
- *experimental simulation* – an experiment that simulates a relevant part of reality under controlled conditions;
- *field experiment* – an experiment conducted in a natural environment, but where the researcher intervenes and manipulates certain factors;
- *field study* – direct observation of a system, with the least possible interference from the researcher;
- *computer simulation* – involves simulating a system or artifact using software;
- *mathematics* – a non-empirical study of abstract structures, their properties, and patterns;
- *logic* – non-empirical reasoning based on sound rules of deduction and argumentation;
- *survey* – a collection of information from a wide and carefully selected variety of stakeholders;
- *in-depth interview* – a structured collection of a (usually larg) amount of information from relatively few individuals;
- *laboratory experiment* – an experiment where the researcher has considerable control and ability to isolate the investigated variables.

I suspect the readers of this book to possess a pretty good understanding of terms like survey, lab experiment, and mathematics, while the other categories may require some additional explanation.

Prototyping is widely used in technology science. When NASA sends vehicles to other celestial bodies, such as the Moon or Mars, they first build prototypes and thoroughly test them in suitable environments on Earth. Prototypes are often simplifications of the final artifact. They can be significantly smaller and built using other materials than the technology they represent.

Experimental simulations are also an effective means of evaluation. An experimental simulation can, for example, involve a wave laboratory in which a new ship hull is tested in different types of sea conditions. Experimental simulation is also popular in human–machine research. It is not unusual to build artificial control rooms to study the behavior of professional operators for different control room designs and setups and different scenarios.

A computer simulation differs from an experimental simulation in that a computer performs it. An experimental simulation often makes use of computer simulation. In an artificial control room, the processes to be controlled by the operators are usually computer simulations of real processes from, for instance, the chemical process industry or energy production.

A field study is about observing without interfering or influencing, for example, how students in a classroom use computer technology in their school work.

Suppose the researcher can manipulate the speed of data communication. In that case, the field study can be turned into a field experiment focusing on how the students change their behavior depending on the speed of the communication network.

Almost all research involves reasoning based on universally valid rules – the kind of reasoning that, to some extent, falls under the concept of common sense, known as logical reasoning. Most research depends on or is based on logical reasoning, even if the arguments are in natural language and not in the form of formal rules that logicians usually work with.

An interview differs from a survey by addressing only a small number of subjects. The questions and answers are, however, more detailed. If the interviews are extensive, they are often referred to as in-depth interviews.

The classification in Figure 5.2 is not complete and can be criticized. It is, for example, difficult to place action research in this picture. The closest we can get is a field experiment. Moreover, what about statistical analysis like machine learning? Some view statistics as a mathematical discipline or a discipline on its own, while others regard it as an integrated part of other research methods.

### 5.2.2 Method triangulation

The term triangulation originates from land surveying and positioning. Knowing the distance to three locations is sufficient to determine where we are in a flat landscape.

Each distance measurement reduces uncertainty about where we are positioned in the landscape. It has inspired the research methodological use of the term triangulation.

> **Definition 5.2** *Method triangulation* implies that certain phenomena are studied from several angles or points of view utilizing different methods.

We can reduce uncertainty by repeating the evaluation:

- with the same layout (positioning and parametrization of instruments and equipment), the sample of subjects, environment, and so on (*Type 1*);
- with the same layout but with other presumed equivalent samples, researchers, input parameters and so on (*Type 2*);
- based on other setups or research methods whose results are expected to correlate with the previous evaluation in a specific way (*Type 3*).

Type 1 triangulation is quite common, especially in the natural sciences. The same evaluation, experiment, or study is repeated time and again based on new measurements or data to get a result that is as accurate as possible.

An example of Type 2 triangulation is repeating a survey on a new set of subjects selected according to the same criteria. If the results diverge, there is cause for concern; if there is compliance, our trust in the findings increases, and we might also be able to improve the accuracy.

Type 3 triangulation may involve comparing results from a computer simulation with physical measurements and thereby calibrating the computer simulation to hopefully give more accurate results in the general case.

### 5.2.3 From the general to the special

Initially, it is problematic in a project and usually not practical to plan the evaluation in detail. We know too little about the invention – the artifact we hope to arrive at or come up with. It is nevertheless advisable at an early stage to carve out an overall evaluation plan that can later be adjusted and specialized as needed. Such a plan should identify the appropriate evaluation methods, the facilities and resources required, and how we envision the practical implementation.

As the invention phase unfolds, it is natural to think through the evaluation plan again, especially the following:

- *Generality:* If we are to claim that the evaluation results hold in a broad context, the evaluation must be organized and conducted in such a way that we can argue for that.
- *Causality:* If we aim to argue that the invention affects its environment, its users, or other phenomena in a cause-effect manner (for example, helping its users in

some way), the evaluation must be planned so that alternative causes can be ruled out.

- *Representativeness:* If we are to model, measure, or estimate (qualitatively or quantitatively) some phenomenon, it must be represented appropriately. For example, a measurement should measure what we intend to measure.
- *Credibility:* If we are to reach convincing conclusions, the evaluation must be performed in such a way that this is feasible; for example, using a suitable statistical method, a well-balanced selection of subjects of the required number, a sufficient number of repetitions, and so on.

In Chapter 11, we introduce four notions of validity corresponding to the four bullets above, namely *external*, *internal*, *construct*, and *conclusion validity*.

## 5.3  Plan for documentation

Testability and repeatability are ideals researchers strive to achieve. The extent to which they succeed depends on themselves, their field of study, and their research method. For example, repeatability is problematic for an action researcher because the action researcher is themself an actor within the object of research, and many factors may affect the outcome. In other disciplines, however, such as the natural sciences, where experiments are conducted in advanced laboratories, a high degree of repeatability is expected.

**Definition 5.3** *Testability* means that other researchers should be able to check our results.

In other words, we have testability to the extent that it is possible for others to inspect and check how we arrived at our findings. This requires proper documentation covering the invention, evaluations, deductions, and data.

**Definition 5.4** *Repeatability* means that other researchers should be able to repeat our studies and experiments.

That is, we have repeatability to the extent that it is possible for others to set up and conduct similar studies and experiments to see if they get the same results. A high degree of testability and repeatability impose stringent documentation requirements that are difficult to satisfy without systematic and careful planning. Important categories of documentation are:

- *invention* – documentation of the new artifact's architecture and design;
- *evaluation setup and procedures* – documentation of how, in what context, and under what conditions the new artifact has been evaluated;
- *data* – documentation of various types of data of relevance for the research;

- *materials* – documentation of various materials of relevance for the research;
- *interpretation and analysis* – documentation of how data and materials have been interpreted, prepared, and processed;
- *deduction* – documentation of arguments and reasoning.

How extensive the documentation plan should be depends on the nature and size of the research project. A plan may be unnecessary if the project involves only one person, and this person is an experienced researcher. On the other hand, in a large international project with many partners, we need plans that reflect all these points. In the following, we look more closely into what this means.

### 5.3.1 Invention

The evaluation aims to substantiate the artifact's different qualities and assess to what extent they fulfill the needs. The artifact must be documented sufficiently to allow others to check, test, or try it.

The nature of the documentation depends on the artifact in question. If the artifact is a building architecture, maybe construction drawings are the suitable format. Moreover, if it is a communication protocol, then so-called sequence diagrams are an option. If it is a chemical substance, then chemical formulas might be a good choice, and if the artifact is a work process or a service procedure, then some flow chart can be used. In the planning phase, choosing a documentation regime suitable for the field in question is essential. Standardized notations and languages should usually be preferred.

### 5.3.2 Evaluation setup and procedures

A prerequisite for repeatability is to accurately record everything from evaluation setup to evaluation results. Suppose the evaluation is a laboratory experiment. In that case, the documentation must contain information about equipment and instruments, how they were configured and parametrized, and how they were shielded from interference and undesirable influences. The procedures, methodologies, and techniques used must also be documented.

Regardless of the type of evaluation, documentation is essential. It should comply with internationally accepted norms for the relevant subject field and type of assessment.

### 5.3.3 Data

Data refers to different representations of information. Data is something we collect, base ourselves on, or results from research activities we perform. It can be anything from registrations and instrument readings to photos, videos, and audio recordings. In a research project, it is common to distinguish between two types of data.

**Definition 5.5**

- *Primary data:* Data collected or generated as part of the project.
- *Secondary data:* Data used by the project but with external origin.

Primary data can thus be understood as new data generated by the project, while secondary data is something we, for example, accessed through libraries. It is also common to distinguish between raw and processed data.

**Definition 5.6**

- *Raw data:* Original, untreated data.
- *Processed data:* Data that has emerged through interpretation and processing.

For an interview, raw data may be a video recording, while processed data is a transcript of the video. The distinction between raw and processed data is relative because processed data from one project can be raw data in another.

How data is documented depends on the type of data. Again, it is crucial to comply with internationally accepted norms.

Requirements capture and many forms of evaluation and analysis involve collecting personal data. Personal data includes opinions, measurements, observations, and various kinds of recordings of human behavior. To the extent that the information is personally identifiable, it falls under data protection law, which is the legal source for privacy protection.

**Definition 5.7** *Privacy* is the ability of an individual or group to seclude themselves or information about themselves and thereby express themselves selectively [91].

The data protection laws vary considerably between different jurisdictions. In the EU, the GDPR (General Data Protection Regulation) specifies general provisions on processing personal data, that is, information that can be directly or indirectly linked to a physical person. Data is classified according to the degree to which individuals are identifiable.

> **Definition 5.8**
>
> - *Anonymized data:* Data where personal identity cannot be derived directly, indirectly, or via a scrambling key.
> - *De-identified data:* Data where identification of individuals is made difficult by representing personal identity with a scrambling key whose link key is stored separately.
> - *Directly identifiable data:* Data where the identity of individuals is clearly stated.

Anonymized data does not fall under the provisions of the GDPR. If de-identification is done correctly, de-identified data is considered anonymous as soon as the link keys are deleted.

There are detailed regulatory frameworks for research data that is not anonymous in many countries. Depending on the amount of data, the type of data, and the degree of the consent of affected individuals, there may be requirements to notify public authorities and obtain authorization.

### 5.3.4 Materials

Materials can be collected, used as a basis, or generated as part of the research. We can document materials by their storage address if they are stored or by their physical properties, for example, a chemical formula. Materials can be classified a bit like data. It can be helpful to distinguish between primary and secondary materials. In archaeological excavation, the artifacts found as part of the excavation are primary, while artifacts extracted from storage to compare with new findings are secondary.

We can also classify materials according to whether they are processed or not. A plant is untreated when picked in the forest but treated when dried and included in a herbarium. The examples above are all from explanation science, but documentation of materials is equally relevant in technology science. A typical example is three-dimensional printing. If we have designed a new kind of three-dimensional printing, examples of prints would be helpful to allow others to compare and measure.

### 5.3.5 Interpretation and analysis

Research involves data processing – manual or electronic. The transition from raw data to processed data involves ethical challenges. De-identification, as we have discussed, is one such. Treatment of so-called *outliers* is another. The selection of data processing methodology is a third.

The order in which things are done is also crucial. How we conduct data analysis depends on the kind of data in question. In hypothesis testing, the hypotheses must be defined before the data is processed. If the data processing is used to identify hypotheses, we need new data to test them. Recordings from interviews will typically be coded and anonymized before further processing. The result will depend on the coding strategy, and this strategy must therefore be documented.

Quantitative data will often be subject to various statistical analyses. The methods used and the assumptions made are crucial to the result and must therefore be carefully described.

### *5.3.6 Deductions*

The reasoning must be sound for the conclusions to be valid. Reasoning can be mathematical reasoning, logical deductions, or more informal inferences based on common sense. In the planning phase, we must decide on the argumentation principles and how they should be made available to others.

# Chapter 6
# Hypotheses

In Chapter 2, we defined a *hypothesis* as an educated guess expressed as an assertion (Definition 2.13). Thus, from a grammatical point of view, a hypothesis is an assertion. When the hypothesis is formulated, it represents a suggestion or idea that we believe can be right or at least fruitful for further investigation. There are many different types of hypotheses. Some are high-level and formulated early in the research process. Others are detailed, postulated late in the invention phase, and exposed to detailed evaluation. Hypotheses of the former category are often referred to as working hypotheses and are revised and improved as the research progresses.

Technology scientists use the concept of hypothesis to a lesser extent than scientists of many other disciplines. In Section 6.1, we try to explain why and that this is not necessarily a problem from a research methodological point of view. Section 6.2 addresses working hypotheses. In particular, we look more closely at hypotheses that assert something about which research method is suitable for reaching a certain goal.

Mature hypotheses exposed to severe evaluation can be classified further. Section 6.3 distinguishes between universal, existential, and statistical hypotheses. This distinction is essential during the evaluation phase.

The philosopher Karl Popper (1902–1994) is widely known for his theory of falsification. Stated briefly: Hypotheses cannot be verified; they can only be falsified. In Section 6.4, we discuss the correctness of this statement. When is it possible to verify, when can we falsify, and when can we strictly do neither?

As elsewhere in the book, we are primarily concerned with technology science, but many technology science projects include sub-projects that belong to explanation science. In this chapter, we, therefore, also consider hypotheses from explanation science.

© The Author(s), under exclusive license to Springer Nature Switzerland AG 2023
K. Stølen, *Technology Research Explained*, https://doi.org/10.1007/978-3-031-25817-6_6

## 6.1 Implicit hypotheses

Much research literature emphasizes the role of the hypothesis. The following adaptation[1] from [27] is a good example:

> An essential aspect of the research process is that the work usually (maybe always?) is related to hypotheses. In some cases, the relationships to hypotheses are easy to spot. The researchers are even aware that they are trying to find support for some hypotheses, disprove other hypotheses, or identify new hypotheses explaining phenomena that cannot yet be explained. In other cases, the hypotheses can be more challenging to spot.

Many technology science publications do not refer to hypotheses at all. In some cases, this is because the authors are weak scientists or writers, but often the hypothesis or hypotheses follow implicitly from other documentation in the publication – in this book, these are referred to as implicit hypotheses, as opposed to explicit hypotheses.

**Definition 6.1** An *implicit hypothesis* is a hypothesis that follows implicitly or can be deduced from other available documentation.

There is an implicit hypothesis as soon as the artifact needs have been described, namely, that *it is possible to build an artifact that satisfies the needs*. Put another way, when the artifact needs have been characterized, we implicitly have a hypothesis of the form:

> It is possible to create an artifact that meets the artifact needs [...].

If we had not believed or at least hoped this was possible, we would hardly have found it worthwhile to describe the artifact needs and initiate the research. The same applies to hypotheses of the form:

> By using the research method [...] it is possible to create an artifact that meets the artifact needs [...].

Hypotheses expressed according to these forms (form since [...] are to be filled in) are implicit. They are also *initial* in the following sense.

**Definition 6.2** An *initial hypothesis* is a hypothesis that can be deduced from the problem analysis (that is, the documentation established as part of the problem analysis).

Contrary to an initial hypothesis, a solution-oriented hypothesis asserts more than what is implicitly deducible. Ideally, this is something creative, innovative, or at least a kind of idea we believe or hope is a step towards a solution.

**Definition 6.3** A *solution-oriented hypothesis* is a hypothesis that claims more than what follows implicitly from the problem analysis.

---

[1] Author's translation with some minor omissions.

Also, solution-oriented hypotheses can follow implicitly from other project documentation, such as a design specification. Since the design specification is a kind of guess or proposal for how to make an artifact that fulfills the artifact needs, we have implicitly:

> It is possible to design an artifact based on the design specification [...] that fulfills the artifact needs [...].

The hypotheses expressed according to this form are implicit but also solution-oriented.

Analogously, when the artifact is completed, the following formats may be used:

> The artifact described in [...] fulfills the design specification [...].
>
> The artifact described in [...] fulfills the artifact needs [...].

In other words, in technology science, hypotheses often follow implicitly from other documentation. Many technology scientists, therefore, do not express their hypotheses explicitly, even though they work in a hypothesis-oriented manner.

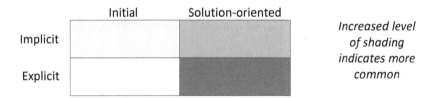

**Fig. 6.1** Relationship between types of hypotheses.

Figure 6.1 summarizes the relationship between the various types of hypotheses introduced in this section. A hypothesis is either initial or solution-oriented. Likewise, all hypotheses are either implicit or explicit.

The box for initial, explicit hypotheses is white because few formulate initial hypotheses explicitly. The box for initial, implicit hypotheses is light gray because such hypotheses occur in all projects with proper problem analysis. However, most hypotheses are solution-oriented; explicit hypotheses are more common than implicit ones, at least if we consider science as a whole (and not just technology science).

In the rest of this chapter, we focus on explicit hypotheses, but the content and the issues covered are equally relevant to implicit hypotheses.

## 6.2  Working hypotheses

During the lifetime of a research project, many hypotheses may appear. Some turn out to be durable, others are further developed or refined, and many are rejected.

The hypotheses will typically change character as the project progresses. Early on, so-called working hypotheses are common.

> **Definition 6.4** A *working hypothesis* is a preliminary hypothesis we accept as a basis for further research.

At an early stage, we may, for example, believe that certain preexisting technologies represent a solution if appropriately combined, which is a working hypothesis. It can be expressed in the form:

It is possible to satisfy the artifact needs [...] by combining the technologies [...].

A hypothesis according to this form is not final. For that, it is too weak. However, it works well as a working hypothesis that is gradually refined or replaced by new hypotheses that assert more on how the technologies should be put together. Such a hypothesis may be in the form:

By combining the technologies [...] according to description [...] the artifact needs [...] will be satisfied.

These more detailed hypotheses are subjected to a thorough evaluation when (and if) we get that far, and they may be suitable for publishing purposes if they prove themselves durable.

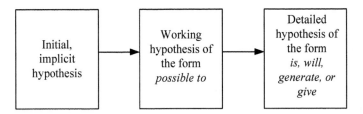

**Fig. 6.2** The role of the working hypothesis.

As illustrated in Figure 6.2, the working hypotheses link the initial, implicit hypotheses and the detailed hypotheses we aim for. Like initial hypotheses, many working hypotheses are weak. Often they only claim that something *is possible*.

The detailed hypotheses are more substantial. They assert that something *is, will, generates, gives*, or the like. In the field of technology science, they tend to describe a recipe for how to build an artifact that meets the needs.

A working hypothesis can also make claims concerning which research method is fruitful or suitable to solve the problem. Such hypotheses we call method-oriented, and many working hypotheses are just that.

> **Definition 6.5** A *method-oriented hypothesis* is a hypothesis that makes some sort of claim about which research method is suitable for solving a particular problem.

The implicit initial hypothesis in Section 6.1 that refers to the research method is method-oriented because it claims its suitability. The research method is described as part of the problem analysis, but usually only at an overall level.

The instantiation occurs later in the project, based on what we think is most fruitful. The suitedness of the chosen research method for the problem in question is a method-oriented hypothesis, although implicit and commonly not referred to as a hypothesis.

Most method-oriented hypotheses occur early in the research process and fall under the concept of a working hypothesis. In some contexts, however, the final hypothesis, subject to detailed evaluation, is method-oriented. A research project may, for example, also contribute to a new or improved research method. In that case, there is a method-oriented hypothesis that requires some evaluation.

*Example 6.1 (Characterizing artifact needs).* Artifact needs may be identified and characterized in various ways. One possible approach is to combine in-depth interviews with surveys. The working hypothesis can then be expressed as follows:

> It is possible to uncover the needs for new technology by conducting a survey followed by in-depth interviews based on the findings from the survey.

When we have worked out plans for both the survey and the in-depth interviews, we can formulate a much stronger hypothesis:

> A survey followed by in-depth interviews according to plan [...] will reveal the needs for new technology.

*Example 6.2 (Use of social networks).* Method-oriented hypotheses are typically weak. They usually only claim that it is *possible* to be successful using the method in question, not that we will be successful independent of how the method is used. The following working hypothesis is a good example:

> It is possible to give a complete picture of youths' use of social networks by conducting in-depth interviews with a small number of youths.

The hypothesis is method-oriented because it proposes a research method: in-depth interviews with youths. It claims it is possible to obtain a complete picture using this method. It is only suited as a working hypothesis, because it does not describe how to conduct the interviews.

One possible detailing of the above hypothesis is the following:

> In-depth interviews, according to `InterviewGuideDoc` of ten youths selected according to the `SelectionProcDoc` provide a complete picture of Norwegian youths' use of social networks.

The latter hypothesis is also method-oriented. However, it is much stronger. It asserts that if we follow the interview guide and select youths according to the procedure, we will get a complete picture.

## 6.3 Universal, existential, and statistical hypotheses

This section explains the differences between universal, existential, and statistical hypotheses. As we will elaborate in Chapters 8–10, the evaluation procedure depends on this distinction. Figure 6.3 illustrates their differences. They are all con-

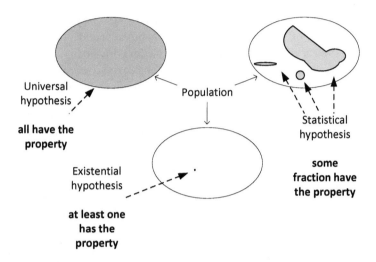

**Fig. 6.3** Illustration of universal, existential, and statistical hypotheses.

cerned with a population. A universal hypothesis asserts that every element or member of the population has or fulfills a particular property. An existential hypothesis asserts that this holds for at least one member of the population. In contrast, a statistical hypothesis, in the simplest case, claims that a particular fraction of the population have this specific characteristic. In the general case, statistical hypotheses describe distributions, but this is not reflected in the figure.

The most straightforward hypotheses address only one population and are either universal, existential, or statistical. More sophisticated hypotheses contain or are composed of sub-hypotheses. They may refer to several populations and are more difficult to classify. We come back to compound hypotheses towards the end of the chapter.

### 6.3.1 Universal hypotheses

A universal hypothesis states that every population member has a specific characteristic. The population can be infinite – for example, all points in time. The population can also be finite but very large. All mammal species that have existed until today is a finite but huge population. The population can also be tiny – for example, the planets in our solar system.

**Definition 6.6** A *universal hypothesis* asserts that every member of a specific population has a particular characteristic.

*Example 6.3 (Examples of universal hypotheses).* The following hypothesis is universal and was among Europeans long considered to be correct:

All swans are white.

The population, in this case, is all swans. It was not until Europeans came to Australia that they discovered that there were indeed black swans.[23]

Many hypotheses of explanation science are universal. A classic example is Newton's first law:

Any object remains at rest or moves at a constant speed in a straight line if no force acts on the object or if the sum of the forces acting on the object is equal to zero.

In this case, the population is all objects. The hypothesis is not correct in general, though, although valid for what was observable in Newton's time. Newton's first law does not apply to tiny bodies like atoms, electrons, and neutrons. For those, the theory of quantum mechanics must be used. Nor does it apply to very high speeds. Then the theory of relativity is needed.

The two universal hypotheses above belong to explanation science. Universal hypotheses are also common in technology science. For Edison's light bulb patent, we can formulate the following universal hypothesis:

Any light bulb built and supplied with a direct current, as specified by Thomas Alva Edison's Patent 223,898 dated January 27, 1880, emits white light.

Universal hypotheses in technology science typically refer to a specification or description of the artifact, as does the patent in the hypothesis above.

In technology science, hypotheses that become subject to a detailed evaluation are typically universal. There are two main reasons. First, as in the hypothesis above, there is a wish to make many copies. It therefore contains a description enabling this in the form of some recipe. Second, the new artifact must be able to handle many different scenarios, alternatively, a high number of parameters or arguments. The following hypothesis, where the artifact is an app, exemplifies the latter.

For any photography of a live animal, the app `AnimalSpecies` identifies the correct species.

---

[2] Black swans were observed for the first time by a European in 1697. It happened in Australia in an expedition led by the Dutch explorer Willem de Vlamingh (1640–1698) [93].

[3] A *black swan* can also signify an extremely rare and unexpected event with significant consequences. The metaphor was in use throughout the seventeenth century and has survived until our days. Today it is perhaps used more than ever due to the writings of Nassim Nicholas Taleb (b. 1960).

In this book, we propose some reusable forms or patterns for hypothesis formulation. They will not be suited to all cases, and the result of filling them in will often sound a bit artificial. However, we may get a helpful starting point for a hypothesis even if the final version or formulation differs slightly.

In the field of explanation science, many universal hypotheses can be written in the form:

**Any** *phenomenon* **of/in** *category* **is/will** [...]

In technology science, the following forms are often helpful:

**Any** *artifact* **built according to** *artifact description* **is/will** [...]

**In any** *context,* **the** *artifact/artifact usage* **generates/provides** [...]

**For any** *argument,* **the** *artifact/artifact usage* **generates/provides** [...]

The terms in italics must be detailed when using the forms. The occurrences of "[...]" are placeholders for the quality, need, or requirements to be satisfied, while "/" distinguishes between different options.

*Example 6.4 (Schematic versions of examples of universal hypotheses).* The use of forms can, as mentioned, give results that sound a bit artificial from a linguistic stand compared to hypotheses that are formulated freely. If we use the forms on the four hypotheses from Example 6.3, we get the following:

**Any** bird **of** any species of swan **is** white.

**Any** object **in** a state of rest or movement in a straight line at a constant speed **will** remain in this state if no force acts on the object, or if the sum of the forces acting on the object is equal to zero.

**Any** light bulb **built according to** Thomas Alva Edison's Patent 223,898 dated January 27 1880 **will** emit white light if supplied with a direct current.

**For any** photography of a living animal, **the** app `AnimalSpecies` **provides** the correct species.

### 6.3.2 Existential hypotheses

An existential hypothesis states that there is at least one element or phenomenon in a population with a specific property. In technology science, the claim might be that it is possible to construct an artifact of a particular kind. An existential hypothesis cannot normally be disproved since the population's size makes it impractical or impossible to examine them all. No matter how long we search, how many phenomena we check, or suggestions for solutions we come up with without finding what we are looking for, a phenomenon that satisfies the hypothesis may nonetheless exist, at least if the hypothesis is not contradictory.

**Definition 6.7** An *existential hypothesis* asserts that a specific population contains at least one specimen with a certain characteristic.

*Example 6.5 (Examples of existential hypotheses).* The abominable snowman is a creature that, according to legends, lives in the Himalayas. The Yeti, as it is called locally, is half-animal, half-human, about 1.80 m high, covered with a reddish-brown coat, with a hairless face [23]. Many have searched for this creature, but none have succeeded in documenting its existence. They have all based themselves on variants of the following existential hypothesis:

> There is an abominable snowman in the Himalayas.

The hypothesis of the abominable snowman constantly reappears, albeit not in serious scientific journals. The reason is that the Himalayas are big, deserted, and not easily accessed. There are, therefore, many opportunities to hide.

In previous chapters, we have seen several similar hypotheses. The hypothesis about the existence of the planet Vulcan (Example 2.5) is a good example:

> There is a planet orbiting in between the Sun and Mercury.

This hypothesis also tends to reappear in new variants. As of 2013, we may, according to [79], exclude the existence of a vulcanoid whose diameter is greater than 5.7 km.

Existential hypotheses are also common in technology science. As explained in Example 2.3, George Cayley defined the concept of the modern airplane as early as 1799, more than 100 years before the Wright brothers constructed a prototype that could fly in a controlled manner. The pioneers of the nineteenth century, inspired by Cayley's concept, based themselves on the following existential hypothesis:

> It is possible to build a powered machine according to Cayley's concept that can carry out controlled flights.

Existential hypotheses also have reusable forms or patterns for hypothesis formulation. Many existential hypotheses in explanation science are expressed according to the following form:

> **There is a** *phenomenon* **of/in** *category* [...]

Likewise, existential hypotheses in technology science can often be expressed along the following lines:

> **It is possible to build an** *artifact* **according to** *artifact description* [...]

*Example 6.6 (Schematic versions of examples of existential hypotheses).* By applying the forms above to the hypotheses from Example 6.5, we get the following:

> **There is a** living specimen **of** the abominable snowman in the Himalayas.
>
> **There is a** celestial body **of** type planet orbiting between the Sun and Mercury.
>
> **It is possible to build a** powered machine **according to** Cayley's concept that can carry out controlled flights.

Existential and universal hypotheses are closely linked. The opposite claim to a universal hypothesis is an existential hypothesis, and the other way around, as exemplified below:

- The opposite claim to the universal hypothesis that *all swans are white* is the existential hypothesis *there is a swan that is not white*.
- The opposite claim to the existential hypothesis that *there are black swans* is the universal hypothesis that *all swans have a color different from black*.

Taking the opposite of something is often called negating.

**Definition 6.8** A claim is a *negation* of another claim if the former expresses the exact opposite of the latter.

Thus, the negation of a hypothesis is a new hypothesis that expresses precisely the opposite of the given hypothesis. If the negation is false, the given hypothesis is true.

*Example 6.7 (Negations of examples of existential hypotheses).* By negating the three existential hypotheses from Example 6.5, we get:

There is no abominable snowman in the Himalayas.

There is no planet orbiting between the Sun and Mercury.

It is impossible to build a powered machine according to Cayley's concept that can carry out controlled flights.

Each negation claims the opposite of the hypothesis it negates. While the hypothesis of the existence of abominable snowmen in the Himalayas at best is very demanding to disprove, its negation is disproved as soon as a specimen is observed. By disproving the negation, we have indirectly verified the original hypothesis that was the starting point. In mathematics, this kind of proof is called *proof by contradiction*.

### 6.3.3 Statistical hypotheses

A statistical hypothesis asserts something about a population as a whole. It often refers to statistical terms such as probability, frequency, or distribution, but this is not what makes it statistical. The hypothesis below is statistical because it claims some average for the population of Norwegian female recruits.

The average height of Norwegian female recruits is 1.65 meters.

Statistical hypotheses are used in most sciences. The following definition, inspired by [98], summarizes what is usually put into the term.

**Definition 6.9** A *statistical hypothesis* is an assertion about the value of one (or more) parameter(s) for a specific population. It can claim something about the parameter's value (such as average, size, or median) or its probability distribution (such as normal, uniform, or logarithmic distribution).

*Example 6.8 (Examples of statistical hypotheses).* We have already seen a statistical hypothesis referring to the average height. The hypotheses below represent other variants:

> The members of the user group are distributed approximately normally for age.
>
> The proportion of women among employees is less than 25%.
>
> The likelihood of a patient dying during a heart transplant is 0.1.
>
> In more than 80% of all Norwegian IT companies with more than 30 employees, implementing the new business architecture will lead to increased profitability.
>
> The average time between subsequent error states for drones based on the new robustness principle is at least 5 hours.

The parameters in the examples above are age, the proportion of women, the likelihood of death during a heart transplant, the proportion of businesses with increased profitability, and the average time between subsequent error states. The populations are represented by user group, employees, heart transplant patients, Norwegian IT companies with more than 30 employees, and drones based on the new robustness principle.

For statistical hypotheses there are also reusable forms that we can employ. In explanation science, the following forms may be helpful:

> *Phenomena* **of/in** *category* **is/will** [...]
>
> **The proportion of** *population* **whose/that** [...] **is/will** [...]

Corresponding forms for technology science are:

> *Artifacts* **built according to** *artifact description* **are/will** [...]
>
> **The proportion of** *population* **whose/that** [...] **when/by** *artifact use* **is/will** [...]

Statistical hypotheses exist in many variants, and there are also other useful forms, but in this book, we are content with these four. Their usage is exemplified in the following.

*Example 6.9 (Schematic versions of examples of statistical hypotheses).* If we use the forms on the hypotheses from Example 6.8, we end up with:

> The age of the members **of** the user group **is** distributed approximately normally.
>
> **The proportion of** employees **whose** gender is female, **is** less than 25%.
>
> **The proportion of** heart transplant patients **that** die during a heart transplant, **is** 10%.
>
> **The proportion of** Norwegian IT companies with more than 30 employees **whose** profitability increases **when** implementing the new enterprise architecture, **is** above 80%.
>
> Drones **built according to** the new robustness principle **will**, on average, operate for at least 5 hours between two subsequent failure states.

The first hypothesis is according to the first form; the second and third hypotheses are according to the second form, the fourth hypothesis is according to the fourth form, and the fifth hypothesis is according to the third form.

## 6.3.4 Compound hypotheses

As mentioned initially in this section, in practice, many hypotheses are compound in the sense that they contain or combine sub-hypotheses. Such hypotheses will typically refer to more than one population. Compound hypotheses can be more difficult to classify as universal, existential, or statistical. The following hypotheses refer to two populations, namely butterfly species, and butterflies of a particular species.

> *H1* – There is a species of butterflies with an average wing span of more than 50 cm.
>
> *H2* – All butterfly species have an average wingspan of less than 50 cm.

Hypothesis *H1* is existential because it claims that at least one butterfly species exists with a specific property. Hypothesis *H2* is universal because it claims something about all butterfly species. What can be confusing is that both contain a statistical sub-hypothesis of the form:

> Specimens of the butterfly species [...] have an average wingspan of [...] than 50 cm.

In other words, *H1* and *H2* can be reformulated as follows:

> *H1* – There is a species of butterflies whose specimens have an average wingspan of more than 50 cm.
>
> *H2* – For every butterfly species, the specimens have an average wingspan of less than 50 cm.

The sub-hypothesis population is butterflies of a particular species in both cases. This population is bound to the butterfly species in question and is therefore subordinate to that. Hypothesis *H1* states that there is at least one butterfly species that satisfies the sub-hypothesis. *H2* states that the sub-hypothesis is true for all butterfly species.

Hypothesis *H3* below is statistical.

> *H3* – At least 10% of all butterfly species are found in Africa.

It concerns the population of butterfly species but contains an existential sub-hypothesis whose population is a particular species of butterflies.

Hypothesis *H4* is also statistical.

> *H4* – At least 25% of all butterflies that occur naturally in Africa, occur only in Africa.

*H4* addresses butterfly species but contains a universal sub-hypothesis concerning butterflies of a particular species.

Hypotheses *H5*, *H6*, *H7*, and *H8* are of the same form as hypotheses *H1*, *H2*, *H3*, and *H4* but are about artifacts instead of butterflies.

> *H5* – It is possible to design a cheap light bulb that, despite daily use, has an average lifespan of at least 40 years.

*H6* – Light bulbs built following the new light bulb principle have an average lifespan of at least 40 years.

*H7* – For at least 80% of all light bulb explosions, turning the light on is a triggering event.

*H8* – Less than 1% of IT-based tools developed for the elderly are used only by the elderly.

These hypotheses are therefore existential, universal, statistical, and statistical, respectively.

The hypotheses above all refer to more than one population but can still be classified as either universal, existential, or statistical. This does not apply to every hypothesis. Hypothesis *H9* consists of two sub-hypotheses, one of which is existential and the other is statistical.

*H9* – All butterfly species have an average wingspan of less than 50 cm, and at least 30% of all butterfly species occur in Africa.

Both sub-hypotheses refer to the population of butterfly species. In our classification, this hypothesis is a hybrid. It can be understood as two independent hypotheses that can be evaluated separately. In practice, hybrid hypotheses can always be decomposed into independent hypotheses that can be assessed individually according to whether they are universal, existential, or statistical. In the following four chapters, all concerned with evaluating hypotheses, we assume for simplicity that every hypothesis is either universal, existential, or statistical.

## 6.4 Can hypotheses be verified?

Suppose we think that abominable snowmen exist. So we have confidence in the following hypothesis:

$H_{YetiExists}$ – There are abominable snowmen.

Suppose we succeed in capturing a being that satisfies the description of an abominable snowman. Have we then verified the hypothesis? In practice, the answer is yes. In theory, it is possible to get involved in philosophical considerations along the lines that we *cannot observe anything with absolute certainty*, but this is not something we will go into here.

The $H_{YetiExists}$ hypothesis is existential. That means, if we find an object or phenomenon that satisfies the description of the hypothesis, it is verified. To falsify an existential hypothesis is, in the general case, impossible. The search space can be infinite; for example, in explanation science, the infinite space surrounding us, or the infinite set of possible designs in technology science. Although we do not find the desired phenomenon where we are searching, or among the technology designs we try out, the sought phenomenon or design may still exist.

When Popper argued that hypotheses could not be verified, he had in mind universal empirical hypotheses of the kind often referred to as laws of nature. A classic example is Einstein's mass-energy law:

$$E = mc^2$$

In natural language, the equation claims that for any physical system, the energy of the system $E$ equals the mass of the system $m$ multiplied by the square of the speed of light in vacuum $c$. Considered a hypothesis, it is universal because it makes this claim for all physical systems. Popper's point is that since there are infinitely many physical systems, and we can only check a finite number of these, the law is impossible to verify. On the other hand, finding one physical system that does not satisfy the law is sufficient to falsify it. This observation does not only apply to so-called laws of nature; it applies to all universal empirical hypotheses that claim something about a multitude of phenomena that are infinite or sufficiently large to make checking them all practical.

Suppose we have established the existence of abominable snowmen and that all the specimens we have examined had four fingers on each hand. We believe that this is true in general and postulate the following hypothesis:

$H_{YetiFing}$ – Abominable snowmen have a maximum of four fingers on each hand.

This hypothesis is not verifiable – only falsifiable. The problem is quite simply that it is not possible to check all abominable snowmen. We can possibly get hold of all the living snowmen, but abominable snowmen may have existed for hundreds of thousands of years, and we cannot go back in time. In other words, we must check all the abominable snowmen that have ever existed to verify $H_{YetiFing}$, while it is sufficient to find one abominable snowman with five fingers on one hand to falsify it.

Based on the discussion above, it is easy to believe that existential empirical hypotheses are verifiable, while universal empirical hypotheses can only be falsified. But the situation is not that simple. First, we can verify a universal hypothesis if the population is sufficiently small and we have full access to check or test out all its members. The hypothesis below can potentially be either verified or falsified:

All the cows in the cowshed of the Norwegian King at Bygdøy on January 1st next year are of the breed Norwegian red fairy.

Verification is impossible only if the population is too large to make it possible or practical. Quite by analogy, the following existential hypothesis can potentially be either verified or falsified:

There is a black cow in the cowshed of the Norwegian King at Bygdøy on January 1st next year.

Second, although in practice we can decide whether something as tangible as an abominable snowman is an abominable snowman if we manage to catch one, in other situations, it is far more difficult to conclude. The following hypotheses illustrate precisely this:

There is biological life in solar systems other than ours.

All biological life in the universe is in our solar system.

The first hypothesis is existential and cannot be falsified; the other is universal and cannot be verified. It is sufficient to find life in another solar system to verify the existential and falsify the universal. As of today, there is no way to acquire a tangible specimen of such a life form unless it comes to us on Earth. In the future, we may be able to obtain indirect evidence via advanced instruments based on complicated theories. To the extent that verification or falsification is relevant, it is under the assumption that these theories are correct.

The hypothesis below is existential but has a universal sub-hypothesis, which again has a statistical sub-hypothesis:

It is possible to construct an artificial human who can solve any task at least as fast and well as at least 60% of all humans.

It cannot be falsified because there are arbitrarily many fundamentally different designs for such an artificial human. Nor can it be verified because there are infinitely many tasks that a human can solve, and it is impossible to test them all.

In practice, terms such as verification and falsification are often too strong. An evaluation will rarely give complete certainty but may increase or decrease our confidence in the correctness of the hypothesis.

Statistical, empirical hypotheses can, in the general case, neither be falsified nor verified.[4] The problem is that we will always lack information. A common hypothesis for six-sided dice used in board games is as follows:

The dice give 6 in $1/6$ of the rolls.

A simple throw does not say anything about the correctness of this hypothesis. If we perform many throws with the same cube, it will probably yield 6 in approximately 1/6 of the cases. However, we can never entirely rule out that this is due to chance, even if the probability of this is tiny.

---

[4] As further detailed in Section 10.2, it is common to argue the correctness of a statistical hypothesis by "falsifying its negation." However, there is no actual falsification – it is only shown that the probability that the result is due to chance or luck is very low.

# Chapter 7
# Predictions

Hypotheses are often evaluated by making predictions about the outcome of tests, studies, or experiments explicitly designed for the purpose. In this chapter, we carefully look at the concept of prediction and its applications in science. We first give a general definition.

> **Definition 7.1** A *prediction* is an assertion about a future condition or circumstance.

A prediction in the general case is thus an assertion of what will happen or what can be experienced, observed, or measured at some future point. This definition is broad. A weather forecast, for example, is a statement about what can be observed in the future, and so are prophecies and divinations. From the Definitions 2.13 and 7.1, it follows that a hypothesis is a prediction to the extent that it claims something about the future, while a prediction is a hypothesis to the extent that it is educated. A weather forecast from a meteorological institute is educated and, therefore, both a prediction and a hypothesis. The claim below, however, is a hypothesis but no prediction since it addresses the past.

The shot that killed the Swedish King Karl XII was fired by a Swedish soldier.

Also, in science, the concept of *prediction* is used with different interpretations (see [75]). In this book, we are concerned with predictions as a means or tool to test or evaluate hypotheses. When we test something, common sense dictates that the test must be related to what is being tested – in our case, the hypothesis. The outcome of the test should potentially be positive or negative for the correctness of the hypothesis. Hence, the outcome of the test must be a consequence of the hypothesis or its opposite in some way or another.

A hypothesis makes an assertion about a population. Predictions forecast the outcome of tests designed to check whether the assertion holds. If we are to test a hypothesis concerning whether a new kind of car engine works as it should, we can build a prototype and try it out. The population is, in this case, car engines of the

K. Stølen, *Technology Research Explained*, https://doi.org/10.1007/978-3-031-25817-6_7

new kind. The alleged quality is that they work as they should. What that means in practice must be specified, but such a specification is irrelevant here.

A prediction will forecast the behavior of the prototype in a concrete test, for example, that the waste gases meet given environmental requirements in a test performed according to an international standard. Other tests with associated predictions will be required to evaluate the prototype for further expectations.

## 7.1 Scientific predictions

There are scientific predictions that are both famous and spectacular without being particularly suitable in an evaluation context. Darwin's prediction about the Madagascan hawk moth is a good example.

*Example 7.1 (Darwin's prediction of moths with extremely long proboscis in Madagascar).* In 1862 Charles Darwin (1809–1882) received a package of orchids for his research on pollination [47]. One of these orchids, known as *Angraecum sesquipedale* and native to Madagascar, surprised Darwin by having a nectary of about 30 cm. The nectar was only excreted on the nectary's lower 3–4 cm. Because this orchid belonged to a group of orchids pollinated by moths, it seemed reasonable to expect this to be the case for *Angraecum sesquipedale*. Hence, Darwin made the following hypothesis:

$H_{Dar}$ – *Angraecum sesquipedale* is pollinated by moths.

Based on this, as well as the design of the nectary, Darwin concluded that in Madagascar, there must be moths with a proboscis that can be up to ten or even eleven inches (25–27.5 cm) [20]. This statement is often referred to as Darwin's prediction about moths. A little simplified, Darwin's prediction can be expressed as follows:

$P_{Dar}$ – In Madagascar, there is a species of moth with a proboscis of at least 25 cm.

The moth proved challenging to find, but in 1903, over 20 years after Darwin's death, Walter Rothschild (1868–1937) and Karl Jordan (1861–1959) [71] published the discovery of a new species of moth native to Madagascar. The moth, aptly called *Xanthopan morganii praedicta*, had a wingspan of about 15 cm and a proboscis of around 30 cm. Later, it was confirmed that this moth pollinates *Angraecum sesquipedale* [87].

Darwin's so-called prediction is problematic from an evaluation perspective. The prediction $P_{Dar}$ does not characterize any test or experiment in the usual sense. It does not specify under what concrete conditions we in *practice* can conclude that it is wrong. On the contrary, $P_{Dar}$ is existential, and as argued in Section 6.4, an existential hypothesis can, in practice, not be falsified when the search space is large. Admittedly, there are at all times only finitely many moths in Madagascar, but to catch and measure all of them is in practice impossible.

To investigate this issue further, we outline an experiment capable of falsifying $H_{Dar}$. We take as starting point $P_{Dar}$ reinforced with the content of Darwin's prediction:

$H_{Dar1}$ – *Angraecum sesquipedale* is pollinated by a moth with a proboscis of at least 25 cm.

The reinforced hypothesis $H_{Dar1}$ implies $H_{Dar}$. If we design an experiment to test the hypothesis $H_{Dar1}$, we implicitly also have a test for $H_{Dar}$.

For simplicity, let us pretend that the species of moth has not yet been discovered and that we, with today's technology, want to perform an experiment that could potentially falsify $H_{Dar1}$. One possible strategy is to monitor a sufficiently large number of orchids around the clock throughout the flowering season so that neither insects nor plants are disturbed. With today's technology, we could install mini cameras recording continuously, day and night. One possible prediction is then:

$P_{Dar1}$ – Either none of the monitored orchids are pollinated, or a moth with a proboscis of at least 25 cm can be observed on at least one of the recordings.

The difference between Darwin's prediction and the above is that the latter forecasts the outcome of a concrete, practically feasible experiment.

## 7.2 Assumptions

Darwin's prediction $P_{Dar}$ does not follow directly from $H_{Dar}$. First, the correctness of the deduction depends implicitly on several facts, such as:

*Angraecum sesquipedale* occurs in Madagascar.

That we implicitly base ourselves on known facts and use facts that are not expressed explicitly is, in practice, necessary. It becomes too tedious and space-consuming to list everything. Only logicians who formally prove deductions need full detail and precision. However, the deduction of $P_{Dar}$ depends not only on implicit facts but also on statements like the one below:

$A_{Dar}$ – Moths must have a proboscis of at least 25 cm to pollinate *Angraecum sesquipedale*.

This statement is not a fact in the sense that it reflects an observation or a finite set of observations. It should instead be understood as a kind of context knowledge whose correctness we have confidence in and therefore assume. The deduction of Darwin's prediction is typical of the relationship between a universal hypothesis and its predictions. Predictions do not follow from the hypothesis alone but from the hypothesis, facts, and assumptions. Only in exceptional cases can a prediction be deducted directly from the hypothesis alone.

When evaluating hypotheses, we almost always need assumptions. We usually rely on surrounding theory if the prediction requires measurement or an experiment. Interesting theories can never be verified once and for all. We must therefore assume their correctness. In practice, this is common and necessary but can lead us astray, as the example below shows.

*Example 7.2 (Assumption about the correctness of Newton's theory of gravity).* As of the discovery of the planet Uranus in 1781, it was clear that the solar system had at least seven planets. The hypothesis that the number of planets in the solar system is seven proved, however, hard to defend. As mentioned in Example 3.3, irregularities were observed in the path of Uranus that broke with Newton's theory of gravity. Based on these observations, it was possible to conclude that either Newton's theory of gravity is wrong or there are other bodies in the solar system that affect the path of Uranus. At the beginning of the nineteenth century, confidence in Newton's laws was almost absolute. It led to the hypothesis that the irregularities were due to another, undiscovered planet, the planet we today know as Neptune.

When Urbain Le Verrier determined path parameters and angular diameter for the undiscovered planet Neptune in 1846, he assumed the correctness of Newton's theory of gravity. He did the same in 1859 when he put forward the hypothesis of the existence of the planet Vulcan with associated predictions of where in space it could be found. This hypothesis was wrong. In the first case, assuming the correctness of Newton's theory was sound and led to the discovery of Neptune. In the second case, however, the same assumption was wrong. The deviation in Mercury's trajectory compared to Newton's theory was not due to an undetected planet but a weakness of Newton's theory. This is a practical dilemma of research. When a prediction fails, the prime suspect is the hypothesis itself or one of the assumptions on which the test or experiment is based. It takes a lot more to question the correctness of a recognized theory.

As indicated above, in addition to assuming the correctness of one or more surrounding theories, we also make assumptions about equipment and the environment. Such assumptions may concern the integrity of data, that instruments are set up correctly, that equipment functions as intended without being affected unexpectedly by external factors, and so on.

*Example 7.3 (Assumptions about data, interview objects, and equipment).* In the aftermath of major accidents and disasters, there are investigations to determine what happened, the causal relationships, and the course of events. The rescue operation will also be subject to rigorous study and assessment. Historical data in the form of logs, pictures, footage, and recordings will be used to test the various hypotheses. The predictions will depend on assumptions concerning the accuracy and the completeness of the collected historical material. The following hypothesis often appears after aviation disasters:

The passenger plane exploded in the air because it was hit by a missile fired from the ground.

Given this hypothesis, we might expect the missile's path to be traceable in historical data, for example, from a military surveillance installation. If we get access to these military data and they disprove the prediction, we have falsified the hypothesis, but only on the assumption that the data was not manipulated before it was passed on to us.

By analogy, if we rely on in-depth interviews of eyewitnesses, experts, relatives, or similar, we must make assumptions about their reliability, communication ability, and honesty. If the weather was good and the plane exploded at a low altitude, we might predict that eyewitnesses to the explosion saw the missile before it hit the aircraft. If the prediction fails, we have falsified the hypothesis, but only under the assumption that the eyewitnesses were honest and the missile was large enough to be observed given the position of the eyewitnesses.

That the equipment works properly is also essential. Using an instrument to test some wreckage depends on assumptions that it works according to specification and that the measurements are not affected by external factors.

## 7.3 Predictions about past events

Predictions are statements or claims about future observations. Predictions may, nevertheless, address events that occurred long before the hypothesis was formulated. Astronomers, for example, make predictions about events that occurred billions of years back in time, and meteorologists test their weather models by predicting past weather based on, for example, glacier drill cores. Such predictions are still of the future since the outcome is first "observed" after the predictions were formulated (even though they could, in principle, have been evaluated long before).

The same also applies to other branches of research. An archaeologist who hypothesizes that a recently excavated settlement dates from the year 500 can predict that a carbon dating performed on organic material found at the settlement will show that the material is about 1500 years old. The prediction could have been tested long before the hypothesis was formulated, but then it would not have served as a prediction in a research methodological sense (at least not if the archaeologist who formulated the hypothesis knew the result).

A similar example is a historian who postulates that a certain medieval prince was secretly in league with the Pope. To test this hypothesis, the historian may predict there exists documentation on this in a particular archive in the Vatican. Again, that the test outcome is not known in advance is essential.

Technology scientists also use predictions that refer to events that have already occurred. If, for instance, the artifact in question is a decision support tool for accident rescue, one prediction might be of the form:

> If the tool is fed with historical data from the already occurred accident [...], the recommendations provided by the tool will be useful to a decision-maker in the following sense [...].

The historical data must be "new" in the sense that the data did not affect or influence the tool's design. Still, the accident from which the data originates may have occurred long before the tool was envisioned.

## 7.4 Reusable forms for technology science

When formulating predictions, reusable forms may be helpful. The frame below offers such a form for each of the ten categories of evaluation methods introduced in Definition 5.1.

**Procedure 7.1** *Reusable forms for predictions:*

*gp01* – *The prototype* **built according to** *artifact description* **will** [...]

*gp02* – **In an experimental simulation based on** *setup*, **the** *prototype* **built according to** *artifact description* **will** [...]

*gp03* – **In a field experiment based on** *setup*, **the** *prototype* **built according to** *artifact description* **will** [...]

*gp04* – **In a field study of** *field description*, **the** *prototype* **built according to** *artifact description* **will** [...]

*gp05* – **Computer simulation of** *reality view* **based on** *setup* **will show that** *artifact description* [...]

*gp06* – **Mathematical reasoning based on** *mathematical theory* **will show that** *artifact description* [...]

*gp07* – **Logical reasoning based on** *logic type* **will imply that** *artifact description* [...]

*gp08* – **A survey based on** *setup* **conducted on** *personnel* **will conclude that** *the prototype* **built according to** *artifact description* [...]

*gp09* – **In-depth interviews based on** *setup* **of** *personnel* **will conclude that** *the prototype* **built according to** *artifact description* [...]

*gp10* – **In a laboratory experiment based on** *setup*, **the** *artifact description* **will** [...]

The terms in italics must be replaced with text. The occurrences of "[...]" are placeholders for descriptions of the artifact needs or requirements the artifact should fulfill.

Evaluations may involve several evaluation methods, often angled towards different parts of the artifact or artifact needs. The terms *prototype* and *artifact description* may therefore refer to only a component of the prototype or an aspect of the artifact description. They should, therefore, be read as *part of the prototype* or *prototype for part of the artifact* and *aspect of artifact description*. The term *setup* is typically a reference to a document that describes a plan or details various parts of the evaluation. Analogously *field description*, *reality view*, *mathematical theory*, *logic type*, and *personnel* represent refinements, points of view, specifications, or selections of the field, the phenomenon, the math, the logic, or the subjects it refers to. A prediction concerns the outcome of a future check, survey, test, experiment, and so on. The forms reflect this through consistent use of the modal auxiliary verb *will*.

The following three chapters exemplify the application of these forms in technology science.

# Chapter 8
# Evaluation of Universal Hypotheses

When our ideas are fruitful and result in some artifact that seems promising, we move on to the evaluation phase. The goal then is to assess if or to what extent the artifact needs have been fulfilled. The basis for the evaluation can be an artifact description, a model, or a prototype, depending on how far we have got in the research process and the selected evaluation method.

We repeatedly evaluate during a research project, and the interval between two evaluations can be short. Some evaluations will be simple and informal, while others are demanding and involve extensive planning. Ideas are evaluated and exposed to various types of analysis as soon as they arise. Some ideas die instantly, while others live for a short while. Only a few are developed to such an extent that they are subject to a detailed evaluation.

The setup of the evaluation depends on the hypothesis in question. Its form and structure will influence the evaluation procedure. This chapter, and the following two, describe procedures for evaluating universal, existential, and statistical hypotheses. Our primary concern is hypotheses that have survived sufficiently long to undergo detailed evaluations.

As explained in Chapter 6, we do not have to express hypotheses explicitly as long as they follow implicitly from other documentation. That the invented artifact satisfies the needs is an example of an implicit hypothesis. We base ourselves on hypothesis-oriented terminology in the following, but everything is equally relevant for implicit hypotheses.

Technology science is closely related to other sciences. There are, therefore, many common features between evaluation in technology science and in science in general. Hence, some of the material is relevant to a much broader domain than technology science.

K. Stølen, *Technology Research Explained*, https://doi.org/10.1007/978-3-031-25817-6_8

## 8.1  Procedure for evaluating universal hypotheses

That the new artifact can be reproduced many times is a common goal in technology science. For example, Edison did not aim at making a single working light bulb but designing a bulb allowing mass production of sufficiently good and robust light bulbs that ordinary people could afford to buy. Another typical expectation is that the new artifact works satisfactorily in various usage situations or for many different arguments or parameter configurations. For example, if the artifact is a computer program, it may be expected to provide correct answers or the correct output values for almost infinitely many arguments or input values. In technology science, the hypotheses are therefore frequently universal.

Universal hypotheses are evaluated by making predictions about the outcome of tests, experiments, and so on. In this respect, technology science does not differ from explanation science. The form of the prediction depends on the chosen evaluation method and the overall context.

A universal hypothesis asserts that every phenomenon of a specific type has certain properties. It is evaluated by testing or checking a finite number of these phenomena for the alleged properties. A prediction in this context is an assertion about the outcome of a test that potentially can falsify the hypothesis. The nature of the test depends on the research discipline and the hypothesis in question.

A test may be to perform an experiment, conduct some investigation, or carry out a case study. Regardless of the test design, it depends on assumptions concerning the correctness of theories, equipment setups, test environments, and potential sources of interference. Predictions will be specialized for the tests. Predictions should be implications of the hypothesis, facts, and the assumptions we make. If a prediction is falsified, it follows from the implication relation that either the hypothesis or an assumption is false (given that we trust the facts).

The concept of implication is thus essential. So let us clarify what an implication is.

> **Definition 8.1** An assertion is an *implication* of (or is implied by) another assertion if it (the former assertion) follows with necessity (from the latter assertion).

*Example 8.1 (Applications of the implication term).* In Example 6.3, we discussed the European sixteenth century hypothesis that all swans are white. Given the fact that there are indeed black swans, the hypothesis is false. Nevertheless, it has many implications, for example, the following three:

> The swans in the garden of the Norwegian Royal Palace are white.
>
> If the bird in the moat is a swan, it is white.
>
> There are no black swans.

It was this last implication that Willem de Vlamingh's expedition disproved as they explored a river in Western Australia and observed black swans on January 10, 1697 [93].

To falsify a universal hypothesis the prediction must be an implication of the hypothesis, the assumptions, and the relevant facts.

**Definition 8.2** A prediction satisfies the *implication requirement for universal hypotheses* if it is an implication of the hypothesis, the assumptions, and the facts together (the conjunction of the hypothesis, the assumptions, and the facts).

Observations consistent with the prediction support the hypothesis. Those conflicting with the prediction give reason to reject the hypothesis or doubt the assumptions made. Hence, if the prediction fails, we have not necessarily falsified the hypothesis because it could be that one of the assumptions is wrong or, in extreme cases, that what we think of as a fact is not a fact after all. On the other hand, we usually have significantly more confidence in the assumptions than in the hypothesis being tested, and even more so in the facts.

The prediction must challenge the hypothesis here and now and not just in principle. It should claim the outcome of a concrete test, experiment, study, or similar.

**Procedure 8.1** *Evaluation of universal hypotheses:*

1. Identify and design several tests (experiments, studies, or similar) suitable to challenge the hypothesis.
2. Formulate one or more falsifiable predictions.
3. For each prediction, check that the implication requirement for universal hypotheses is satisfied for the hypothesis and the assumptions made.

If (at least) one prediction is falsified, either the hypothesis is wrong, or at least one assumption does not reflect reality. Otherwise, we may increase our confidence in the hypothesis and the outcome of the tests can be used to argue for its correctness.

We can test an interesting hypothesis in many ways – often infinitely many. We can only perform a small number of these. The tests must therefore be chosen carefully so that the hypothesis is challenged as much as possible within the available budget and context.

Figure 8.1 illustrates Procedure 8.1 with the help of a class diagram. The boxes and lines represent concepts and relationships, respectively. The arrowheads specify the reading directions. The expressions $*$ (finitely many) and $1\ldots*$ (from one to finitely many – that is, not zero, and not infinitely many) characterize the number of occurrences (multiplicity). In natural language: An evaluation of a universal hypothesis involves at least one test. The conjunction of the universal hypothesis, the assumptions, and the facts imply a nonzero number of falsifiable predictions. Each prediction makes a claim about the outcome of a test.

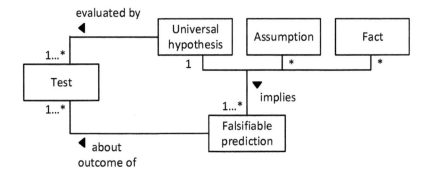

**Fig. 8.1** Evaluation of universal hypotheses.

## 8.2 Examples

In the following, we use Procedure 8.1 in ten different examples. We apply Procedure 7.1 to formulate predictions. Each form is used at least once. Hypotheses are formulated using the forms introduced in Section 6.3.1. They are summarized below.

**Procedure 8.2** *Reusable forms for universal hypotheses in technology science:*

    *guh1* – **Any** *artifact* **built according to** *artifact description* **is/will** [...]

    *guh2* – **In any** *context*, **the** *artifact/artifact usage* **generates/provides** [...]

    *guh3* – **For any** *argument*, **the** *artifact/artifact usage* **generates/provides** [...]

The terms in italics must be replaced by text. The occurrences of "[...]" are placeholders for descriptions of the artifact needs or requirements that should be satisfied.

### 8.2.1 Prediction for prototyping

Prototyping is about making an example or instance of a design or concept, often simplified, for testing and experimentation. In technology science, prototyping, often in combination with other evaluation methods, is common. We can prototype the invention entirely or focus on a part or aspect of it, for example, a particular component.

Prototyping often gives rise to predictions in the form:

*gp01 – The prototype* **built according to** *artifact description* **will** [ . . . ]

Here we refer to the concrete prototype we have built or obtained access to and that we will make use of in the evaluation.

*Example 8.2 (Prototyping of Edison's light bulb design).* For Edison's variant of the light bulb from Example 6.3, the form *guh1* may be filled in as follows:

**Any** light bulb **built according to** Thomas Alva Edison's Patent 223,898 dated January 27, 1880, **will** emit white light if provided with a direct current.

Since the hypothesis is universal, it is a prerequisite that the patent is sufficiently detailed to guarantee the claimed outcome for any correct realization. The hypothesis addresses only one aspect of the artifact needs: the white light.

We test the hypothesis by making a bulb prototype following the patent and then checking whether this works correctly. In practice, we must restrict ourselves to a set of voltage levels selected according to a test plan. The prediction below is expressed according to form *gp01*.

The light bulb, **built according to** Thomas Alva Edison's Patent 233,898 dated January 27, 1880, **will** emit a white light if provided with a direct current, according to TestPlanDoc.

If the prediction is falsified for at least one voltage level, we have falsified the hypothesis, given that the prototype is built correctly according to the patent. Otherwise, we have an indication that the hypothesis is correct.

## *8.2.2 Prediction for experimental simulation*

As we have mentioned before, the ten categories of evaluation methods are partially overlapping. An experimental simulation may involve some form of prototyping but will be more than a prototyping because the environment of the prototype is simulated.

Using experimental simulation to evaluate universal hypotheses often yields predictions in the form:

*gp02 –* **In an experimental simulation based on** *setup,* **the** *prototype* **built according to** *artifact description* **will** [ . . . ]

The replaceable term *setup* is necessary because a prediction will refer to a plan or specification for setting up and conducting the experimental simulation.

*Example 8.3 (Experimental simulation for evaluation of smoke-diving robot).* Making the best use of smart technology in a crisis is an active field of research. Suppose a group of scientists have developed a smart (or more "intelligent") robot for smoke diving and postulated the following hypothesis (based on *guh1*):

> **Any** smoke diver robot **built according to** `SpecSmartRobotDoc` **will**, on average, locate fire victims faster than existing robots.

To test this hypothesis, we can perform an experimental simulation where an old industrial plant is equipped to simulate a fire in a chemical plant. One possible prediction expressed according to *gp02* is then:

> **In an experimental simulation based on** igniting and smoke screening an industrial plant, as described in `SimulationPlanDoc`, **the** prototype **built according to** `SpecSmart-RobotDoc` **will**, on average, locate fire victims faster than [. . . ]

The placeholder "[. . . ]" is replaced with references to existing robots that our robot should improve on.

### 8.2.3 Prediction for field experiment

As the invention matures, it will be tested in its intended environment. For this purpose, field experiments may be a good choice. Again, the artifact is often represented by a prototype. A field experiment differs from an experimental simulation in that the environment is genuine and not simulated. As the term suggests, a field experiment involves some form of experimentation. The researchers will manipulate a small number of parameters, either in the environment or in the setup of the artifact itself. In summary, in a field experiment, the researchers will first and foremost observe, but they may adjust and try different settings during the experiment.

The prediction form is analogous to the one for experimental simulation.

> *gp03* – **In a field experiment based on** *setup*, **the** *prototype* **built according to** *artifact description* **will** [. . . ]

*Example 8.4 (Field experiment to evaluate the scalability of a service architecture).* Many Internet service providers struggle to meet scalability requirements. Service scalability is the ability of the service to continue fulfilling user needs as the number of concurrent users increases. For example, scalability is a problem in the case of extraordinary events. Many have experienced that messages sent on special occasions like New Year or during crisis-like events have been delayed.

Suppose a group of researchers has developed a new IT architecture to improve the scalability of Internet services. Assume further that their hypothesis is as follows:

> **For any** Internet service, **the** re-implementation based on the new service architecture **provides** improved scalability.

This hypothesis is formulated according to *guh3*. In practice, the hypothesis will be limited to services of a particular type and, in addition, refer to a description or specification of the new architecture. Still, for simplicity, this is ignored here.

Such a hypothesis can be evaluated in different ways, depending on how far we have gotten in the research process and our confidence in the hypothesis. If the hypothesis survives long enough, it will typically be tested in a field experiment. Such a field experiment may include re-implementing an existing service based on the new architecture. The two implementations of the same service, the old one and the one based on the new architecture, can both be offered online to service users. The users may not be aware of the ongoing field experiment, but the researchers may compare the two implementations to see how they scale with increased usage. The prediction they test may be something like:

> **In a field experiment based on** users being distributed equally at all times between the two implementations, **the** implementation **built according to** the new service architecture **will** scale better than the old implementation.

The term *setup* corresponds to "users being distributed equally at all times between the two implementations."

## 8.2.4 Prediction for field study

In a field study, the artifact is used or integrated into its natural or intended environment, and the role of the researchers is limited to observation. So a field study is similar to a field experiment except that the interference of the researchers is very restricted. The predictions are highly dependent upon the scope of the study, for example, the object and concept of observation and the period in question. In the form below, the *field description* is a placeholder for a description of this scope.

> *gp04* – **In a field study of** *field description*, **the** *prototype* **built according to** *artifact description* **will** [ ... ]

*Example 8.5 (Field study for a web-based cooperation approach).* Institutions and businesses consisting of many units that function pretty well individually, but struggle with inter-unit collaboration, are known as silo-oriented organizations. Suppose a group of researchers has come up with a new web-based approach known as IntCoop to improve inter-unit collaboration within silo-oriented organizations. Assume their hypothesis is:

> **In any** business organized according to the silo principle, **the** introduction of IntCoop **provides** better inter-unit cooperation.

The hypothesis is universal and follows *guh2*, but in practice, unrealistic without further characterization of assumptions and context. However, this is ignored here.

We intend to evaluate the hypothesis in a field study. It takes place at the institution BigOrg, whose organizational structure is silo-oriented. Based on introductory studies, the researchers concluded that IntCoop requires a training and start-up phase of four weeks to work properly. Based on that, the researchers have come up with the following prediction:

> **In a field study of** daily operations at IntCoop for eight weeks, **the** prototype **built according to** IntCoop **will** meet the expectations of improved cooperation after a training and start-up phase of four weeks.

In practice, the prediction will refer to documents further detailing the field description and the expectations of the outcome.

### 8.2.5 Prediction for computer simulation

Computer simulation is widely used to check or estimate the properties of artifacts. A pure computer simulation means that computer programs model the artifact and its intended environment. The computer simulation consists in executing these programs. We may infer, decide, or forecast the artifact's properties based on the results.

Computer simulation is often used with other evaluation methods but may also be used as a stand-alone method. In the latter case, the following form may be helpful:

> *gp05* – **Computer simulation of** *reality view* **based on** *setup* **will show that** *artifact description* [...]

The phrase *prototype* **built according to** that appeared in previous forms is less relevant here as the choice of the method implies that both the artifact and the relevant part of its intended environment are computer programs.

*Example 8.6 (Computer simulation of air resistance).* An artifact will usually undergo much evaluation before we get as far as trying it out in a field experiment or a field study – at least if the artifact is expensive to produce. Suppose the invention is a new form of surface treatment for aircraft hulls. If so, the costs associated with making a realistic prototype are substantial. It will, therefore, hardly be built without extensive prior evaluation. In this context, a computer simulation is a valuable tool.

Consider the following hypothesis expressed according to *guh3*:

> **For any** conventional aircraft hull, **the** surface treatment, according to ProcedureDoc **provides** a reduction in air resistance of at least $X\%$.

The hypothesis is universal for the population of conventional aircraft hulls. For this book, what is meant by *conventional aircraft hulls* is not essential. $X$ is a constant – for example, the digit 5.

Suppose the scientists have created a simulation program, Simul, to predict the air resistance for different wind directions and atmospheric conditions, detailed in SimulationSetupDoc. A possible prediction is then:

> **Computer simulation of** aerodynamics **based on** SimulationSetupDoc **will show that** ProcedureDoc reduces the air resistance of the conventional aircraft hull [...] by at least $X$ %.

The placeholder "[...]" should be filled in with the type of conventional aircraft hull that is simulated. We get a prediction of the form above for each type of conventional aircraft hull considered.

## 8.2.6 Prediction for mathematics

Mathematics, as we will deal with in the following, and logic, which we will return to in the next section, are non-empirical evaluation methods. In this respect, they differ from the eight other method categories in Definition 5.1. Concerning infinitely large populations of physical phenomena, universal hypotheses can never be verified, while existential ones can never be falsified. Such restrictions do not necessarily apply to mathematics and logic.

Mathematics is a foundation for all kinds of science and a necessary ingredient in many evaluations. Mathematics is, for example, often required to arrive at testable predictions for a hypothesis.

When using mathematics as a stand-alone method, predictions of the following form may be helpful:

> *gp06* – **Mathematical reasoning based on** *mathematical theory* **will show that** *artifact description* [...]

There are many different kinds of mathematics, and *mathematical theory* is a placeholder for a specification of the theory on which the prediction is based.

*Example 8.7 (Use of mathematics for testing of an algorithm).* Many mathematical functions can be tough to calculate precisely (for some, if not all) values. In that case, it might be a good idea to come up with a computer program that calculates the function or at least provides an estimate with some accuracy. Suppose our researchers have made such a computer program EstF for some specific function $f$, and that their hypothesis is as follows:

> **For any** argument $a$, **the** program EstF **generates** a result whose maximum deviation from $f(a)$ is less than $X\%$.

Again we have used *guh3*, and again $X$ is a constant. Although the function can be too time-consuming or even impossible to calculate manually in the general case, a trained mathematician may be able to determine or at least estimate the result with some accuracy for a few arguments. Assume InputListDoc lists those arguments. In that case, we may predict:

> **Mathematical reasoning based on** conventional arithmetic **will show that** EstF for all arguments in InputListDoc yields a result whose deviation from $f(x)$ is less than $X\%$.

Instead of testing the computer program for various arguments, we can try to prove its correctness. In that case, the following prediction is relevant:

**Mathematical reasoning based on** conventional arithmetic **will show that** EstF for any argument yields a result whose deviation from $f(x)$ is less than $X\%$.

The two predictions are, by nature, different. To falsify the first, it is sufficient to show that EstF deviates from the calculated value of $f$ by at least $X\%$ for at least one argument in InputListDoc.

The same is also sufficient to falsify the second prediction, but the hypothesis and the second prediction can be false even if the first prediction is true. The reason is that EstF can satisfy the accuracy requirement for the arguments in InputListDoc, but fail for at least one of the (possibly infinitely many) other arguments.

If, on the other hand, we succeed in proving that EstF calculates $f$ with a deviation of less than $X\%$, and we have great confidence in the correctness of the proof, we have, in practice, verified the hypothesis. In theory, we can never entirely rule out the possibility of a mistake, independent of how thoroughly we check, but in practice, we will consider the hypothesis correct.

### 8.2.7 Prediction for logical reasoning

We use logical argumentation all the time, in science and daily life, although we do not necessarily refer to it as such. Logical reasoning is the principle we base ourselves on when we put together arguments to substantiate a claim. These principles fall to a certain extent under the category of common sense. Logical argumentation is an ingredient in any evaluation. It is the glue that binds the evidence together.

In this section, however, our concern is logical reasoning as a stand-alone method. In this context, predictions in the following form may be helpful:

*gp07* – **Logical reasoning based on** *logic type* **will imply that** *artifact description* [...]

*Example 8.8 (Logical argumentation to evaluate a requirement specification).* We have previously argued that working out a requirement specification is an early task in many projects. The requirements specification is a step towards the invention the project is aiming for. However, there are also research projects where the creation is a requirement specification – in other words, a requirement specification is the artifact to be delivered to the client.

As an example of such a project, assume a public agency has commissioned a group of researchers to come up with a detailed requirements specification for a particular kind of life jacket. The researchers are, of course, constrained by the public agency in various ways. In particular, there are standards and general guidelines that must be taken into account. There are also limits on manufacturing costs. For simplicity, let us refer to the limitations laid down by the client as GivenConstrDoc. Assume further that the research group has arrived at a draft specification that we simply refer to as ReqDoc that they believe fulfills the limitations imposed by the public agency. Based on *guh1* we get the following hypothesis:

**Any** implementation **built according to** `ReqDoc` **will** satisfy `GivenConstrDoc`.

To test this hypothesis, we can use logical reasoning. Assume that `RecDoc` and `GivenConstrDoc` can be formalized in the predicate logic. In that case, we can predict:

**Logical reasoning based on** predicate logic **will imply that** `Rec2Doc` satisfies `Given-ConstrDoc`.

Predicate logic is a formal tool that logicians use to reason about statements. Computer tools can check logical arguments automatically, but human intuition and time-consuming manual work may be required in the general case. In practice, the predictions will be weaker, for example, focusing on some of the limitations and only specific requirements.

### 8.2.8 Prediction for survey

Surveys are in widespread use. Common application domains are societal problems, human opinions, or human needs and desires. Surveys are used in marketing to provide knowledge as a basis for sales strategies and in politics to determine public opinions. Opening a newspaper without finding references to surveys in one form or another is hardly possible. In technology science, surveys are used, among other things, to gather user experiences, map artifact needs, and, not least, evaluate to what extent expectations have been met. Here we focus on the latter, and in such a context, the following form may be helpful:

*gp08* – **A survey based on** *setup* **conducted on** *personnel* **will conclude that** *the prototype* **built according to** *artifact description* [...]

*Example 8.9 (User interface evaluation survey).* Suppose a group of researchers has come up with a new user interface for smartphones specially adapted for the elderly. Suppose further that `NewInterfaceDesignDoc` describes this interface, and that the hypothesis (based on *guh2*) is as follows:

**In any** context where the elderly are dependent upon a smartphone, **the** interface based on `NewInterfaceDesignDoc` **provides** improved performance.

One way to test the hypothesis is to allow a group of potential users, `Selected-ElderlyDoc`, to try out prototypes built according to the design description. User experiences can be collected by employing a survey based on a suitable questionnaire. A possible prediction is:

**A survey based on** `SurveySetupDoc` **conducted on** `SelectedElderlyDoc` **will conclude that** smartphones with a user interface **built according to** `NewInterface-DesignDoc` satisfy `Expectations2PerformanceDoc`.

The document `SurveySetupDoc` describes the context, execution, selection of elderly subjects, questionnaire, and so on. The document `Expectations2-PerformanceDoc` characterizes what it means for the survey results to satisfy the expectations for improved performance.

### 8.2.9 Prediction for in-depth interview

In-depth interviews of users or subjects participating in a trial of some prototype, for instance, a field experiment or an experimental simulation, is a common procedure when evaluating new artifacts. For this purpose, the form below is relevant:

> *gp09 –* **In-depth interviews based on** *setup* **of** *personnel* **will conclude that** the *prototype* **built according to** *artifact description* [. . . ]

*Example 8.10 (In-depth interview on the health effects of a mattress).* Suppose a group of scientists has developed a new kind of mattress based on a material that is supposed to make the mattress better suited for people with a specific type of back disorder. Their hypothesis could be something like this:

> **For any** patient with the given back disorder, **the** new type of mattress **provides** improved quality of life without compromising on other essential features such as lying comfort, durability, and fire safety.

This hypothesis based on *guh3* exemplifies the need to combine different evaluation methods. To test the effect of the mattress on back disorder, a clinical study might be a good choice. On the other hand, fire safety can be tested separately in a laboratory. Suppose we have access to several users with the relevant back disorder who have tried out a prototype of the new mattress over time, and that we will conduct in-depth interviews with a selection of them. The prediction might then be expressed along the following lines:

> **In-depth interviews based on** `InterviewGuideDoc` **of** `SelectedMattressUsers-Doc` **will conclude that** the mattresses **built according to** a design based on the new material meet user expectations.

### 8.2.10 Prediction for laboratory experiment

A laboratory may be used both for the invention phase and for the evaluations. A laboratory makes it easier to control the environment and factors that could affect the outcome. Many predictions aiming at evaluation in a laboratory can be expressed as follows:

> *gp10 –* **In a laboratory experiment based on** *setup,* **the** *artifact description* **will** [. . . ]

*Example 8.11 (Laboratory experiment for testing of an avalanche probe).* In the case of snow avalanches, avalanche probes are used to search for potential victims. Suppose a group of researchers has developed a new avalanche probe design based on material that makes the rod slip through the snow more easily. Their hypothesis (based on *guh3*) is as follows:

> **For any** type of snow, **the** avalanche probes built with the new material **generate** less friction than conventional avalanche probes.

A laboratory makes it easier to control the surroundings and factors that could affect the outcome. It makes good sense to combine different evaluation methods. To test the hypothesis in a realistic setting, we may use experienced avalanche rescue personnel to try out the new avalanche probes during avalanche rescue exercises. We can collect their experiences using in-depth interviews, possibly supplemented by a survey. However, friction can also be measured in a laboratory, and in such a context, the following prediction might be relevant:

> **In a laboratory experiment based on** `SetupFrictionExsperimentDoc`, **the** avalanche probes built with the new material **will** give less friction than conventional avalanche probes for the following kinds of snow `SnowTypeSpecDoc`.

# Chapter 9
# Evaluation of Existential Hypotheses

Existential hypotheses are most common early in the research process. They mostly appear as working hypotheses that are subsequently refined or reformulated in universal or statistical form. Occasionally, however, also existential hypotheses are exposed to detailed evaluations. In this chapter, we explain how to proceed in such cases.

At the very end, we also elaborate on the extent to which a detailed evaluation of a refined or further developed hypothesis can be understood as an evaluation of the working hypothesis that was its starting point.

## 9.1 Procedure for the evaluation of existential hypotheses

A common and rather convincing approach to arguing the correctness of a hypothesis asserting the existence of a particular phenomenon or that a specific artifact can be realized is to try to find or construct a specimen. Since *one* copy is enough, the implication requirement from the universal case is reversed. We still insist on implication, but the hypothesis and the prediction swap places.

> **Definition 9.1** A prediction satisfies the *implication requirement for existential hypotheses* if the hypothesis is an implication of the prediction, the assumptions, and the facts together (the conjunction of the prediction, the assumptions, and the facts).

In other words, the prediction is required to imply the hypothesis given the assumptions and the facts.

**Procedure 9.1** *Evaluation of existential hypotheses:*

1. Identify and design tests (experiments, studies, and similar) suited for supporting the hypothesis.
2. Formulate one or more verifiable predictions.
3. For each prediction, check that the implication requirement for existential hypotheses is satisfied for the hypothesis and the assumptions made.

If (at least) one prediction is verified, the hypothesis is correct, or at least one of the assumptions does not reflect reality. The confidence in the hypothesis increases to the extent the assumptions are realistic, and the outcome of the tests can be used to argue for this increase. Otherwise, the confidence in the hypothesis is reduced.

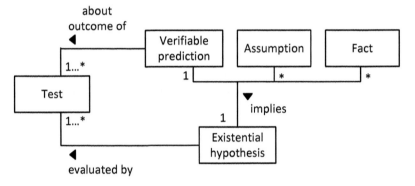

**Fig. 9.1** Evaluation of existential hypotheses.

As summarized in Figure 9.1, the tests are designed to evaluate the hypothesis. A prediction makes a claim about the outcome of a test. The prediction, assumptions, and facts together imply the hypothesis.

*Example 9.1 (Evaluation of Darwin's prediction).* Although the $P_{Dar}$ of Example 7.1 is known as Darwin's prediction, it also constitutes a hypothesis – an existential one. One possible evaluation strategy is to equip an expedition to Madagascar to look for moths whose proboscises are at least 25 cm. Suppose that `SearchPlanDoc` describes the exploration plan. In that case, we can predict:

$P_{Ex}$ – At least one moth caught when implementing `SearchPlanDoc` will have a proboscis of 25 cm or more.

If $P_{Ex}$ is verified, $P_{Dar}$ is also verified. In the opposite case, we have an indication that moths with a proboscis of at least 25 cm are nonexistent in Madagascar. The strength of the indication depends on the thoroughness of the exploration.

The negation of an existential hypothesis is universal, as explained in Section 6.3.2. An existential hypothesis can therefore also be verified by falsifying its negation.

**Procedure 9.2** *Indirect evaluation of existential hypotheses:* Evaluating an existential hypothesis indirectly involves evaluating its negation using Procedure 8.1. If the negation is falsified, we have verified the hypothesis that was negated.

Whether we evaluate an existential hypothesis directly or indirectly is a question of what is most convenient and, in some cases, personal preferences. Below we redo Example 9.1 using the procedure for indirect evaluation.

*Example 9.2 (Indirect evaluation of Darwin's prediction).* Negating $P_{Dar}$ gives:

$P_{DarNeg}$ – All moths in Madagascar have a proboscis that is shorter than 25 cm.

To falsify a claim that something does not exist, we must find a specimen of this something or at least establish clear indications that this something exists. Given the expedition described in Example 9.1, we may predict:

$P_{ExNeg}$ – All moths caught when collecting moths according to `SearchPlanDoc` will have a proboscis shorter than 25 cm.

If $P_{ExNeg}$ is falsified, $P_{DarNeg}$ is also falsified, and then $P_{Dar}$ is verified. In the opposite case, we have indications that such moths do not exist in Madagascar and can support an argument for this. The solidity of the argument will once more depend on the thoroughness of exploration.

## 9.2 Examples

The use of Procedure 9.1 can be combined with generic forms. In Section 6.3.2, we proposed the following.

**Procedure 9.3** *Reusable forms for existential hypotheses in technology science:*

*geh* – **It is possible to build an** *artifact* **according to** *artifact description* [...]

The terms in italics are replaced by text when using the form. The occurrence of "[...]" is the placeholder for the artifact needs or requirements to be satisfied.

Predictions may be expressed as in the universal case using Procedure 7.1. Since evaluating an existential hypothesis typically involves building a prototype, the six forms explicitly referring to prototyping are especially useful. The remaining four can also be used, but they correspond to more specialized or theoretical studies based on specifications and design drawings before going into the often costly step of building a realistic prototype.

*Example 9.3 (Evaluation of Cayley's concept).* Cayley's concept for constructing a powered airplane (see Example 6.5) illustrates the challenges associated with evaluating existential hypotheses. Cayley's proposal corresponds to an existential hypothesis of the form:

> **It is possible to build a** powered machine **according to** Cayley's concept that can carry out controlled flights.

Cayley published his concept in 1799. In 1903, the Wright brothers undertook the first controlled flight with a plane based on Cayley's concept. Hence, it took more than 100 years before Cayley's hypothesis was verified.

The prediction below is based on *gp02*. The aircraft constructed by the Wright brothers is the prototype.

> **In an experimental simulation based on** Experiment Setup, **the** aircraft of the Wright brothers **built according to** Cayley's concept **will** perform a controlled flight.

The alternative hypothesis below is weaker.

> **It is possible to build a** powered machine **according to** Cayley's concept that can take off by its own engine and land safely.

The corresponding prediction is based on *gp01*.

> The aircraft of the Wright brothers **built according to** Cayley's concept **will** take off by its own engine and land safely.

Note that although the forms from Procedure 7.1 can be used regardless of whether the hypotheses are universal or existential, the processes are different. In the universal case, a falsified prediction means the hypothesis is false, given that the assumptions hold and that the prototype is built according to the artifact description as specified in the hypothesis. In the existential case, we can only conclude that the prototype in question is insufficient. On the other hand, if the prediction is verified, the hypothesis is verified in the existential case, while we have a positive indication in the universal case at best.

Mathematics is non-empirical and, therefore, different from the natural and social sciences. However, if the notion of an artifact is interpreted broadly, mathematical proofs may be understood as artifacts, and proving mathematical theorems is the same as inventing artifacts. A famous *artifact* in this respect is the proof of Fermat's last theorem. The search for such a proof turned out to be even more demanding than building airplanes based on the concept of Cayley, as explained below.

*Example 9.4 (Fermat's last theorem).* Fermat's last theorem states that for any integer $n$ greater than 2, it is impossible to find positive integers $a$, $b$, and $c$, so that

$$a^n + b^n = c^n$$

This theorem was first postulated in 1637 by the French lawyer and mathematician Pierre de Fermat (1601–1665) in the margin of a copy of *Arithmetica*[1] [77]. Fermat claimed that he had a proof but that it was too large to fit on the same page. This comment inspired numerous mathematicians to try to come up with a proof. It was not until 1995 that someone succeeded.

The mathematicians who all failed in the 358 years that passed from Fermat's comment until Andrew Wiles (b. 1953) published his proof, all based themselves on the following existential hypothesis:

> **It is possible to build a** mathematical proof for the theorem **according to** Fermat's postulation in the copy of *Arithmetica*.

Has Wiles's proof verified the hypothesis once and for all? It may be discussed. Only a few mathematicians in the world, in addition to Wiles, understand and can check the proof in full. We cannot entirely rule out that someone will find a mistake sometime in the future, even though the belief in the correctness of the proof is firm. Mathematicians are thus somewhat in the same situation as technology scientists. Although we succeeded in inventing an artifact that seems to fulfill the relevant needs, we may have overlooked issues or weaknesses that may appear later.

As we have seen, we can evaluate an existential hypothesis by repeatedly attempting to construct prototypes based on the principle, idea, or concept specified in the hypothesis. These prototypes are subsequently checked or tested. We have verified the hypothesis if we succeed in making a prototype with suitable properties. Otherwise, we may continue until we have exhausted our budget or lost confidence in the hypothesis. Since an idea may be realized in numerous ways, it is far from certain that we will succeed, even if the hypothesis is correct.

## 9.3 Working hypotheses and evaluation

Working hypotheses are subject to tests or trials as part of the invention phase. Before reaching the evaluation phase, they have, in most cases, been further developed or refined. However, the evaluation of these improved or more detailed hypotheses can also shed light on the correctness of the working hypotheses. This requires that the evaluated hypothesis and the working hypothesis are sufficiently related. In the following, we establish what this means.

---

[1] A mathematical text written by the Greek mathematician Diophantus (ca. 210–ca. 295).

In a technology-scientific context, working hypotheses are usually existential. Let WorkHyp denote an existential working hypothesis. The more detailed hypothesis that undergoes a detailed evaluation is either universal, existential, or statistical. Let UniHyp, ExHyp, and StatHyp denote hypotheses for each option. We assume they are related to WorkHyp as specified in Table 9.1. If these relationships do not hold, the outline below is irrelevant.

**Table 9.1** Relationship between detailed and working hypothesis

| Detailed hypothesis | Relation to WorkHyp |
|---|---|
| UniHyp | implies |
| ExHyp | implies |
| StatHyp | implies with a high likelihood |

The implications in Table 9.1 will, in practice, depend on several assumptions, which are ignored in the following.

Since UniHyp implies WorkHyp, a positive outcome for UniHyp is also a positive outcome for WorkHyp. On the other hand, WorkHyp may be true though UniHyp is false, which means that a negative result for UniHyp does not say much about the correctness of WorkHyp.

*Example 9.5 (Immunity to prostate cancer).* Suppose we have a theory that for genetic reasons, some humans cannot get a specific type of cancer, such as prostate cancer. The existential working hypothesis can then be:

> There is a genetic code that makes men immune to prostate cancer.

Assume further that we have received funding for a research project based on this hypothesis and that we at some point identify a genetic code that we suspect has the desired property. If so, we would like to evaluate the following hypothesis:

> Men whose DNA satisfies GenCodeDoc will not develop prostate cancer.

This hypothesis is universal. The domain is men whose DNA has the relevant code. The universal hypothesis implies the working hypothesis, given that there are men with such DNA. If the universal hypothesis is falsified, we have excluded one genetic code, but there are many more, and the working hypothesis may still be correct. On the other hand, a favorable outcome is correspondingly positive for the working hypothesis.

Since ExHyp implies WorkHyp, it follows that a positive evaluation for ExHyp is also favorable for WorkHyp. On the other hand, a negative evaluation for ExHyp is not necessarily negative for WorkHyp, as ExHyp can be significantly stronger than WorkHyp. It may, however, give some indication that WorkHyp is wrong.

*Example 9.6 (Detailing the hypothesis of the Higgs particle).* In 1964 six physicists published a mathematical model that explained why elementary particles have mass. This model implied the existence of a hitherto undiscovered particle, later referred to as Higgs particle, after Peter Higgs (b. 1929), one of these physicists.

In 2012, based on new knowledge and contributions from other physicists, the hypothesis for the existence of the Higgs particle could be expressed as follows [68]:

$H_{Higgs}$ – There is an undetected particle with a mass between 114 and 140 GeV.[2]

$H_{Higgs}$ substantially reinforces the original hypothesis, which does not refer to mass (except that it is not zero). We can therefore think of the original hypothesis as a working hypothesis implied by the more detailed hypothesis. Both are existential. Verifying the more detailed hypothesis is also a verification of the original working hypothesis.

In March 2013, CERN announced that they had empirical evidence for the existence of such a particle [13]: "[...] the new particle is looking more and more like a Higgs boson [...]". The mass of the new particle was around 125 GeV.

The prediction they tested was in the form:

$P_{Higgs}$ – In an experiment where the particle accelerator in CERN is configured according to the following setup [...] an undetected particle with a mass between 114 and 140 GeV will be observed.

It is based on a large number of assumptions about the surrounding theory as well as the setup of the experiment.

The statistical case is analogous to the universal, although the implication requirement has a probabilistic element.

*Example 9.7 (Computer program for the board game Go).* Writing a computer program that could play Go better than any human was long seen as virtually impossible. This changed in March 2016, when AlphaGo beat the world champion, Lee Sedol (b. 1983), 4–1 in a game over five sets. Although they knew it was highly challenging, the developers of AlphaGo started with a working hypothesis along the following lines:

It is possible to create a computer program that plays Go better than professional Go players.

One hypothesis they could have postulated to test AlphaGo is:

AlphaGo wins, on average, at least 90% of all games against top-ranked professional Go players.

This hypothesis is statistical. It implies the existential working hypothesis with high probability. If the evaluation of the statistical hypothesis is positive, the confidence in the working hypothesis increases accordingly.

---

[2] The measuring unit GeV stands for gigaelectronvolt.

The following table summarizes the discussion above.

**Table 9.2** Consequence for the working hypothesis of the evaluation outcome

| Detailed hypothesis | Evaluation outcome | Consequence for the working hypothesis |
|---|---|---|
| UniHyp | positive | positive |
| | negative | none |
| ExHyp | positive | positive |
| | negative | negative indication |
| StatHyp | positive | positive |
| | negative | none |

In all three cases, a positive evaluation outcome is also positive for the working hypothesis. A negative outcome gives a negative indication only in the existential case.

# Chapter 10
# Evaluation of Statistical Hypotheses

In this chapter, we address the evaluation of statistical hypotheses, emphasizing statistical hypothesis testing. We first explain what statistical hypothesis testing is. Then the actual evaluation procedure is introduced and, after that, exemplified. As before, we use generic forms to formulate hypotheses and predictions.

The procedures for universal, existential, and statistical hypotheses must be combined to evaluate compound hypotheses (see Section 6.3.4). At the very end of this chapter, we explain what this means in practice.

This book give a broad coverage of its subject. This chapter does not require prior knowledge of statistics or mathematics.

## 10.1 Brief introduction to statistical hypothesis testing

Statistical hypothesis testing is well described in many textbooks. Our goal here is not to compete with these textbooks but to clarify the relationship between hypotheses and predictions in statistical hypothesis testing. To this end, we must introduce some basic concepts. The underlying math is, however, not needed and entirely omitted.

Statistical hypothesis testing is about building an argument for the truth of a statistical hypothesis by arguing that its negation, the opposite statement, most likely is false. The negation of the hypothesis is called the null hypothesis. The prefix *null* reflects that it represents the null state.

**Definition 10.1** A *null hypothesis* characterizes the null state, for example, the current situation, the prevailing perception, or what is possible with already existing solutions or technology.

K. Stølen, *Technology Research Explained*, https://doi.org/10.1007/978-3-031-25817-6_10

The alternative hypothesis is the hypothesis we are trying to build evidence for.[1] It represents an alternative to the prevailing practice, view, or current theory.

**Definition 10.2** An *alternative hypothesis* characterizes the alternative to the null hypothesis, the alternative we are trying to argue for.

*Example 10.1 (Alternative and null hypothesis for a new medicine).* Suppose we have developed a new drug for a human disease for which 60% are cured using medications already on the market. We expect the new medicine to be better, meaning more than 60% get cured. Hence, the alternative hypothesis is as follows:

$H_{Alt}$ – The healing rate for the new medicine is higher than 60%.

The null hypothesis expresses precisely the opposite:

$H_{Null}$ – The healing rate for the new medicine is less than or equal to 60%.

A statistical hypothesis claims something about a population. The population is a collection of whatever kind. It may be humans with a particular disease, as in the example above, houses built in Greece in the 1950s, one-eyed humans, or blue stones. Typically a population is large – often infinitely large. The null hypothesis is evaluated by predicting under what conditions it should be rejected based on data collected from a (usually small) selection of the population in question.

Erroneously rejecting the null hypothesis (known as a Type 1 error) is considered worse than keeping the null hypothesis when the alternative hypothesis is true (known as a Type 2 error). Potential costs associated with introducing new products or revising theories and principles that depend on the null hypothesis mean that we want to be on the safe side before rejecting the null hypothesis. In other words, the probability of incorrectly rejecting the null hypothesis must be small.

Significance levels characterize an upper limit on the magnitude of this probability. It is common to operate with a significance level, or statistical significance, of 5% (0.05, expressed as a probability). If the consequence of a Type 1 error is considerable, the requirement for significance is stronger – for example, less than or equal to 1%.

## 10.2 Procedure for the evaluation of statistical hypotheses

Evaluating the alternative hypothesis by testing the null hypothesis is often "incorrectly" referred to as trying to falsify the null hypothesis. Strictly speaking, since we only have data from a small sample, the null hypothesis cannot be falsified. In the same way as we may win a fortune in a lottery, we can be "lucky" and get statistical

---

[1] In practice, we may postulate several alternative hypotheses, but here for simplicity, we limit ourselves to one.

significance due to "chance" even though the null hypothesis is correct. However, the likelihood of this happening is small. The implication requirement for statistical (null) hypotheses reflects this.

**Definition 10.3** A prediction satisfies the *implication requirement for statistical hypotheses* if the null hypothesis, the assumptions, and the facts together imply that the probability of the prediction being "wrongly" falsified is less than or equal to the significance level.

If the prediction is falsified, the probability of a Type 1 error is less than or equal to the significance level.

**Procedure 10.1** *Evaluation of statistical hypotheses:*

1. Formulate the null hypothesis by negating the alternative hypothesis.
2. Select the significance level.
3. Formulate a falsifiable prediction that satisfies the implication requirement for statistical hypotheses.

If the assumptions hold and the prediction is falsified, one of the following applies:

- The null hypothesis is false, and the alternative hypothesis is true.
- The null hypothesis is true, and the prediction was falsified by chance although the probability for this was less than or equal to the selected significance level.

If the assumptions hold and the prediction is not falsified, one of the following applies:

- The null hypothesis is true, and the alternative hypothesis is false.
- The null hypothesis is false, and the alternative hypothesis is true, but we have not found evidence for this.

As summarized in Figure 10.1, the null hypothesis negates the alternative hypothesis we hope to find evidence for. The null hypothesis, assumptions, and facts imply the prediction for the selected significance level. The prediction makes a forecast considering the data collected from a finite subset of the population.

## 10.3 Examples

Procedure 10.1 requires the formulation of hypotheses as well as predictions. In the following, we express hypotheses using the forms introduced in Section 6.3.3. They are summarized below.

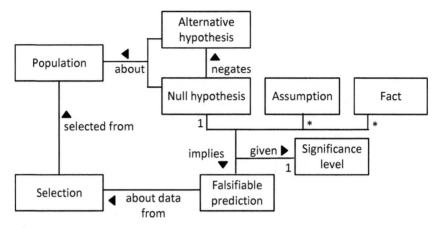

reading direction of relation
1    one occurrence
*    arbitrarily many occurrences

**Fig. 10.1** Evaluation of statistical hypotheses.

---

**Procedure 10.2** *Reusable forms for statistical hypotheses in technology science:*

   *gsh1* – *Artifacts* **built according to** *artifact description* **are/will** [. . . ]

   *gsh2* – **The proportion of** *population* **whose/that** [. . . ] **when/by** *artifact use* **is/will** [. . . ]

The terms in italics are replaced by text when using the form. The occurrence of "[. . . ]" is the placeholder for the artifact needs or requirements to be satisfied.

---

We use the forms from Procedure 7.1 to express predictions. For statistical hypothesis testing, the forms *gp01*, *gp02*, *gp03*, *gp05*, *gp08*, and *gp10* are the most useful. The forms *gp04* and *gp09* are less relevant since field studies and in-depth interviews usually are qualitative. The forms *gp06* and *gp07* may be applied for non-empirical statistical hypotheses in statistics or probabilistic logic.

*Example 10.2 (Prediction for null hypothesis for a new medicine).* The hypotheses $H_{Alt}$ and $H_{Null}$ from Example 10.1 reformulated according to *gsh2* give:

   $H_{Alt'}$ – **The proportion of** humans with the disease [. . . ] **that** is cured **by** the use of the new medicine **is** greater than 60%.

   $H_{Null'}$ – **The proportion of** humans with the disease [. . . ] **that** is cured **by** the use of the new medicine **is** less than or equal to 60%.

The population is humans with the disease in question. Suppose this is a common disease, which means that the population is vast. We can usually test the new medicine only on a small selection – let us say 20 people. In other words, based on the data we get by treating these 20 persons with the new medicine, we hope to be able to conclude something about the correctness of $H_{Alt'}$.

The challenge is to formulate a prediction $P$ whose falsification implies that the likelihood of $H_{Null'}$ being wrong is very high. Curing 60% of 20 means that 12 will become healthy. If 13 become healthy, we have cured 65%. Suppose that the significance level is 5%. In that case, at least 17 must be cured. With the help of *gp03*, we get the following prediction:

> **In a field experiment based on** `ExSetUp`, **the** medicine **built according to** `Chemical-Formula` **will** cure a maximum of 16 out of the 20 patients with the disease [...].

As mentioned earlier, the forms are primarily for making outlines that then can be reformulated more intuitively. The prediction above may, for example, be simplified to:

> In an experiment based on `ExSetUp`, the medicine `ChemicalFormula` will cure a maximum of 16 out of the 20 patients with the disease [...].

If the prediction is falsified, the probability of $H_{Null'}$ being true is less than 0.05, implying that the probability of $H_{Alt'}$ is greater than or equal to 0.95 $(1 - 0.05)$.

*Example 10.3 (Experimental simulation of alarm management).* Being able to cope with many alarms triggered almost simultaneously is a significant problem in many control rooms. A large number of simultaneous alarms is almost as dangerous as a lack of alarms or false alarms. When an engine of an Airbus A380 aircraft belonging to Quantas exploded in the air in November 2010, 54 different alarms were triggered simultaneously. This was a much higher number of simultaneous alarms than the pilots were trained to handle [12]. In critical situations, alarms must be triggered and presented so that the operators can make the right decisions.

Suppose a group of researchers has invented a new principle for presenting alarms to operators. There are many different control rooms, and the same solution is not suited for all. Some control rooms are small and have only one operator, while others are large and controlled by a team of operators. In addition, there are significant differences between operating a chemical process and, for example, police operations in a major city. Let us assume that our researchers are concerned with a kind of control room that we, for simplicity, call `ClassC`. Two possible hypotheses (using *gsh1* and *gsh2*) are:

> Alarm management systems **built according to** the new alarm management principle **will** reduce the likelihood of fatal operator mistakes in `ClassC` control rooms.
>
> **The proportion of** operator decisions concerning alarms in `ClassC` control rooms **that** are erroneous **when** using the new alarm management principle **is** less than 0.1%.

The corresponding null hypotheses are:

> Alarm management systems **built according to** the new alarm management principle **will** not reduce the likelihood of fatal operator mistakes in `ClassC` control rooms.
>
> **The proportion of** operator decisions concerning alarms in `ClassC` control rooms **that** are erroneous **when** using the new alarm management principle **is** equal to or greater than 0.1%.

To test the null hypotheses, we can, for example, perform experimental simulations using an artificial control room that, for the occasion, is equipped following the `ClassC` standard. The processes to be controlled can be computer simulations – in other words, software programs mimicking the behavior of the actual processes. The artificial control room should have various devices for monitoring operator behavior, documentation, and experimentation. Examples of such are cameras, eye-tracking sensors, sound sensors, and so on. The idea is to measure the ability of operators to handle different scenarios depending on how the alarms are communicated and presented. The operators may, for example, be hired from a chemical plant operated via a `ClassC` control room. Possible predictions for the two hypotheses are then:

> **In an experimental simulation based on** `ExPlan1Doc` in the artificial `ClassC` control room, **the** alarm management system **built according to** the new alarm management principle **will** give at least $X\%$ fatal operator errors.

> **In an experimental simulation based on** `ExPlan2Doc` in the artificial `ClassC` control room, **the** alarm management system **built according to** the new alarm management principle **will** give at least $Y$ operator errors per $Z$ operator decisions.

$X$, $Y$, and $Z$ are constants that must be carefully determined given the relevant context and the requirements for significance.

*Example 10.4 (Multiple uses of forms for statistical hypothesis testing).* Below we offer a statistical (alternative) hypothesis for each of the Examples 8.2, 8.3, 8.4, 8.6, 8.9, and 8.11.

- Light bulbs **built according to** Thomas Alva Edison's patent 223,898 **will**, on average, have a lifespan of at least 100 hours.
- Smoke-diving robots **built according to** `SpecSmartRobotDoc` **will** have an average operating time of at least 5 hours between subsequent failure states.
- **The proportion of** Internet services **whose** scalability is improved **by** re-design based on the new service architecture **will** be at least 80%.
- **The proportion of** commercial flights **whose** flight time is reduced by at least 10% **when** the aircraft hull is treated according to `ProcedureDoc` **will** be at least 90%.
- **The proportion of** elderly people **whose** experience of mastery increases **when** using an interface based on `NewInterfaceDesignDoc` **will** be greater than 60%.
- Avalanche probes **built according to** the new material design **will**, on average, function satisfactorily for at least 96 hours.

The corresponding null hypotheses (the negations of the alternative hypotheses) are as follows:

- Light bulbs **built according to** Thomas Alva Edison's patent 223,898 **will** on average have a lifespan of less than 100 hours.
- Smoke-diving robots **built according to** `SpecSmartRobotDoc` **will** have an average operating time of less than 5 hours between subsequent failure states.

- **The proportion of** Internet services **whose** scalability is improved **by** re-design based on the new service architecture **will** be less than 80%.
- **The proportion of** commercial flights **whose** flight time is reduced by at least 10% **when** the aircraft hull is treated according to `ProcedureDoc` **will** be less than 90%.
- **The proportion of** elderly people **whose** experience of mastery increases **when** using an interface based on `NewInterfaceDesignDoc` **will** be less than or equal to 60%.
- Avalanche probes **built according to** the new material design **will**, on average, function satisfactorily for less than 96 hours.

Below is a prediction for each null hypothesis according to Procedure 7.1.

- A selection of $X$ light bulbs **built according to** Thomas Alva Edison's Patent 223,898 **will**, on average, function for less than $Y$ hours.
- **In an experimental simulation based on** the ignition and smoke screening of an industrial plant, as described in `SimulationPlanDoc`, **the** prototype **built according to** `SpecSmartRobotDoc` **will** have an average operating time of less than $X$ hours between subsequent failure states.
- **In a field experiment based on** users being distributed equally at all times between two implementations of the same Internet service, **the** implementation **built according to** the new service architecture **will** scale better than the same Internet service built according to the old service architecture for fewer than $Y$ of a selection of $X$ Internet services.
- **Computer simulation of** the relationship between flight time and aerodynamics for a sample of $X$ commercial flights **based on** `SimulationSetUpDoc` **will show that** surface treatment of aircraft hulls according to `ProcedureDoc` reduces the flight time by at least 10% for fewer than $Y$ flights.
- **A survey based on** `SurveySetupDoc` **conducted on** `SelectedElderly-Doc` **will conclude that** the user interface **built according to** `NewInterface-DesignDoc` increases the experience of mastery for fewer than $X$ elderly people.
- **In a laboratory experiment based on** `ExSetUpDoc`, where avalanche probes are exposed to stress and wear corresponding to active search, **the** $X$ avalanche probes built with the new material **will**, on average, function satisfactory for less than $Y$ hours.

The constants $(X, Y)$ must be carefully selected for each prediction.

## 10.4   What if the hypothesis to be evaluated is compound?

Hypotheses can be compound, as we saw in Section 6.3.4. In the following, we demonstrate the use of Procedures 8.1, 9.1, and 10.1 to guide the evaluation of compound hypotheses. Consider once more the compound hypothesis from Section 6.4:

> It is possible to construct an artificial human who can solve any task at least as fast and well as at least 60% of all humans.

This hypothesis is existential but includes a universal sub-hypothesis, which again contains a statistical sub-hypothesis. The population of the existential hypothesis is all possible (potential designs of) artificial humans, and the population of the universal hypothesis is all possible tasks. In contrast, the population of the statistical hypothesis is all human beings.

Getting from the compound hypothesis to a detailed prediction using the introduced procedures involves three steps.

1. *Establish a prediction that satisfies the implication requirement for existential hypotheses.* Suppose that after much effort, we have invented a robot design `ArtHumanDesign` that we want to test. Suppose further that `ArtHuman` is a prototype built according to this design. The following prediction based on form *gp01* satisfies the implication requirement for existential hypotheses:

   > `ArtHuman` **built according to** `ArtHumanDesign` **will** solve any task at least as fast and well as at least 60% of all humans.

   The prediction is universal since it makes a claim for every task. It is compound since it contains the same statistical sub-hypothesis as the compound hypothesis above. It makes a prediction since it asserts something about the future. This claim, however, is not suited in its current form for a detailed evaluation since we cannot test infinitely many tasks. In our research methodological context, it may therefore be understood as another hypothesis for which we must identify an evaluation strategy.

2. *For the prediction from Step 1, establish a prediction that satisfies the implication requirement for universal hypotheses.* One possible evaluation strategy is to perform a finite number of experimental simulations, each aimed at a specific task. There are infinitely many tasks to select from, like solving a crossword puzzle, directing traffic, or stacking wood logs. Choosing tasks we suspect will be challenging for the robot seems reasonable. We then get a prediction according to form *gp02*:

   > **In an experimental simulation based on** `SimulationPlanDoc`, **the** `ArtHuman` **built according to** `ArtHumanDesign` **will** solve each task in `TaskSpecDoc` at least as fast and well as at least 60% of all humans.

   This prediction is statistical, but still not suited for a detailed evaluation. We may think of it as an alternative hypothesis intended for statistical hypothesis testing.

3. *For the prediction from Step 2, establish a prediction that satisfies the implication requirement for statistical hypotheses.* The prediction from Step 2 gives the following null hypothesis:

> **In an experimental simulation based on** `SimulationPlanDoc`, **the** `ArtHuman`
> **built according to** `ArtHumanDesign` **will** solve each task in `TaskSpecDoc` at least
> as fast and well as fewer than 60% of all humans.

Given a sample of 20 humans and a significance level of 5%, we get the following
prediction:

> **In an experimental simulation based on** `SimulationPlanDoc`, **the** `ArtHuman`
> **built according to** `ArtHumanDesign` **will** solve at least one task in `TaskSpecDoc`
> at least as fast and well as fewer than 17 humans from the sample of 20.

If the prediction is verified, there is at least one task for which we lack sufficient
evidence to reject the null hypothesis. In that case, we must reject the universal prediction resulting from Step 2. The existential compound hypothesis that
was our starting point has not been falsified, but we have established that `Art-`
`HumanDesign` probably does not represent a solution.

On the other hand, if all the predictions resulting from Step 3 are falsified, we
may argue that the alternative hypotheses most likely are true and that `Art-`
`HumanDesign` meets the expectations based on the evaluations conducted so
far.

The procedure exemplified above is summarized in a generalized form below.

**Procedure 10.3** *Evaluation of compound hypotheses:*

1. Apply Procedures 8.1, 9.1, or 10.1 on the compound hypothesis depending on
   whether it is universal, existential, or statistical.
2. Repeat Step 1 on the resulting prediction until it is not compound.
3. Apply Procedures 8.1, 9.1, or 10.1 on the prediction resulting from Step 2
   depending on whether it is universal, existential, or statistical.
4. Evaluate the prediction resulting from Step 3.
5. Given the evaluation outcome of Step 4, deduce the implications for the compound hypothesis from Step 1 by backtracking the steps needed to get to the
   prediction from Step 3.

# Chapter 11
# Quality Assurance

Conducting evaluations is demanding, and mistakes are easily made. Quality assurance is therefore essential. An evaluation should ideally be *valid* and *reliable*. In the following, we explain what this means in practice. We first define the relevant concepts at an overall level. After that, these are broken down into sub-concepts.

**Definition 11.1** An *evaluation* is *valid* if it evaluates what it is meant to evaluate.

If an evaluation is valid, we should get approximately the same result each time it is repeated. This is known as reliability. Detailed documentation so that the evaluation can be replicated by others is a prerequisite for reliability.

**Definition 11.2** An *evaluation* is reliable if it can be repeated and gives approximately the same result each time it is repeated.

An evaluation can be reliable without being valid. Suppose we are testing a new drug on humans. Moreover, assume we are unaware of the placebo effect, namely that some patients experience a benefit from inactive "look-alike" medicine or treatment. If so, we may observe a repeatable positive effect, although the drug does not work. In that case, the evaluation is reliable but not valid since we are not capturing the drug's actual effect. In other words, we have reliability but not validity.

On the other hand, if an evaluation is valid, it is also reliable. Validity means that the test is set up so that we observe the drug's actual effect. In that case, the test is also reliable as we will necessarily get the same result each time it is repeated.

In practice, validity and reliability are unattainable ideals. Instead, we try to estimate the degree of validity and reliability. Reliability is particularly problematic in the social sciences. Action research where the researcher and the object of research are not separated is an example.

In the following, we decompose the concepts of validity and reliability into sub-concepts.

## 11.1 Validity

It is common to decompose validity into four sub-concepts, namely external validity, internal validity, construct validity, and conclusion validity. In the following, we address these concepts one by one and identify potential threats and sources of error.

### *11.1.1 External validity*

External validity aims to capture the relevance of the evaluation beyond the specific context in which it was carried out. Are the results as general as the researchers claim? Do they provide knowledge relevant to objects, individuals, phenomena, and artifacts other than the one(s) involved or made use of? Are the results of value in different situations or contexts, at other times, and under other conditions?

> **Definition 11.3** An *evaluation* is *externally valid* if the domain, the situation, the context, and the time aspect for which it is alleged to apply or be relevant are correctly characterized.

*Example 11.1 (Lack of external validity).* Suppose we have developed a new method for security testing of web applications. Suppose further that we have completed an experiment where:

- Half of the subjects used our method, while the other half used a market-leading method.
- Each subject worked individually and had the same time available.
- Both groups tested the same web applications.
- The group that used our method scored better with statistical significance.

Since we tend to have faith in our work, we may easily become over-eager and put forward the findings as more general than they are. For example, we might claim that the results hold for security testers, web applications, and security threats in general. This may, however, conflict with external validity unless the test subjects were truly representative of security testers. For example, if we used students as test subjects, the generalization to security testers would hardly be externally valid.

Another issue is the monetary and administrative cost of experiments with many human subjects. To reduce costs, we are often forced to minimize the time spent by each subject. This is problematic for external validity because web applications that are sufficiently simple to be tested in a short experiment hardly represent web applications in an industrial context. Analogously, the number of security threats and scenarios we can cover is limited.

A third issue is the rapid evolution of Internet technologies and Internet-based solutions. What is secure today is not necessarily secure tomorrow. In other words, though our security-testing method is better than the market-leading methods today, the market-leading methods could be more appropriate for future technologies.

In short, external validity addresses the degree to which the scope of relevance is correctly characterized. Possible sources of error include:

- *Subjects:* The selection of human subjects considering their competence, age, and experience. Some academics make too-strong claims based on experiments with students as test subjects.
- *Artifacts:* The selection of things, processes, and structures. Suppose we perform an experimental simulation: Does the layout match reality, and is it as general as we claim?
- *Location:* The selected location(s) may also impact the result. If we are to measure the effect of acid rainfall in forests,[1] a forest next to a river to which calcite is added artificially is hardly suitable.
- *Time:* People's perceptions and society are constantly changing. The outcome of a survey may be influenced by events that occurred in the last few days before it took place.

## 11.1.2 Internal validity

An evaluation is internally valid if the causal relationships it claims to have established are genuine and complete.

> **Definition 11.4** A *relationship* between two events $A$ and $B$ is *causal* if
>
> - there is a correlation between $A$ and $B$ in the sense that $B$ occurs every time $A$ occurs;
> - $A$ always occurs before $B$ in time – that is, $A$ and $B$ are strictly arranged in time;
> - there is no plausible alternative explanation for the correlation between $A$ and $B$.

For example, if we conclude that a specific health effect is due to medication based on some new substance, this is a claim of a causal relationship. Internal validity requires the alleged causal relationships to be authentic.

Internal validity is irrelevant for purely descriptive evaluations, meaning they only document observations without identifying causal relationships.

> **Definition 11.5** An *evaluation* is *internally valid* if the causal relations that the evaluation claims to have established are real.

---

[1] Research results on the effect of acid rainfall on forests were taken very seriously in the 1980s by the political leadership in Europe. It is an excellent example of research whose validity was overrated by its contemporaries [70].

*Example 11.2 (Lack of internal validity).* We consider once more the security-testing method from Example 11.1. Internal validity requires that our method score better due to its qualities. In other words, the cause of the claimed effect was the new method, not hidden factors or interference during the experiment. For example, if we did not consider the subjects' background and expertise when setting up the groups, one group could, on average, be more competent than the other. If so, this skewed distribution of expertise could have given better scores for our method. Furthermore, if the market-leading method uses charts with colors, and we did not consider the extent to which the subjects were color blind, this could be another reason why the market-leading tool scores less well.

There are many threats to internal validity. Some of the most important ones are presented below:

- *Arrangement in time:* Definition 11.4 requires that the cause *A* occurs before the effect *B*. In some cases, cause and effect may be mixed. Events may have a mutually reinforcing effect so that more of one gives more of the other. If increased travel leads to increased use of social networks and vice versa, what is the cause, and what is the effect?
- *Third variable:* Research literature often refers to the cause as the independent variable and the effect as the dependent variable due to its dependency on the cause. A common source of error is the existence of a hidden cause in the form of a third variable closely associated with the incorrectly identified cause. A good example is a positive correlation between the number of churches and the number of criminal incidents in a city. More churches do not lead to more crime, but the number of churches is influenced by a third variable, namely the size of the city.
- *Distribution:* As indicated in Example 11.2, the distribution of subjects between groups quickly leads to distortions and wrong conclusions. This applies to humans and generally to allocating animals, plants, artifacts, stone fragments, etc., between groups or quantities. For example, suppose we are to measure the effect of different fertilizer programs on barley. In that case, the seed grain must be of approximately the same quality for the various programs that are being tested.
- *External event:* Subjects, including laboratory animals, may easily be affected by external circumstances. If we are trying to measure to what extent a meditation technique is soothing, a natural disaster in the subjects' vicinity during the experiment could easily affect the result.
- *Maturation:* Living beings often change their behavior as they age or mature. This can make measurement difficult and lead to errors in a study that continues over a long period. For example, a child's natural maturation can be misinterpreted as an effect of a particular type of treatment.
- *Repeated testing:* When the same or a similar test is repeated several times, the learning effect may result in measurement errors. If we repeatedly expose the same human to intelligence tests of the same kind, the score will usually increase, although the intelligence remains the same.

- *Method impact:* The instrument, form, or, more generally, the research method we use can affect the result. Subjects can unconsciously change their ways of thinking. The instrument's smell, sound, or color can affect animals' behavior. In a dark pub, more light will make it easier to play billiards. At the micro-level, turning on a lamp can ruin the whole experiment.
- *Confirmation errors:* Scientists can also be influenced to find what they believe is right. It need not be conscious[2] but can still lead to fundamental errors. Gregor Johann Mendel (1822–1884), who discovered the laws of inheritance, was accused after his death of getting "too-good results". This has led to extensive literature right up to our days [67]. In Mendel's case, the problem is not that problematic since he was right. An example with far graver consequences is Prosper-René Blondlot's (1849–1930) discovery of N-rays in 1903, which proved to be an illusion [64]. Blondlot did not only fool himself. In fact, from 1903 until 1906, more than 100 educated scientists published articles on N-rays and their properties.

Internal validity is often problematic and especially difficult in evaluations involving human beings. To argue for internal validity, we must exclude the existence of alternative causes, for example, by systematically considering each potential threat in the list above.

## 11.1.3 Construct validity

Construct validity concerns the correspondence between theoretical concepts, constructs, and relationships appearing in the hypothesis and the theory we build on, and their representation or operationalization in the evaluation setup.

> **Definition 11.6** An *evaluation* is *construct valid* if abstract concepts, constructs, and relationships from surrounding theories, hypotheses, and assumptions are correctly operationalized in the form of measurement points, indicators, and metrics.

*Example 11.3 (Lack of construct validity).* Consider once more Example 11.1 and the following hypothesis:

> Security testers are more effective when using our method than the market-leading method.

We assumed the results were positive and that the users of our method, on average, scored better with statistical significance. But what does that mean for the truth of the hypothesis? The answer depends on the degree to which the scoring metric measures the effectiveness of security testing correctly. There are many potential pitfalls. For example, if each test subject has only two hours to complete the test,

---

[2] But there are also many examples of deliberate cheating; see, for example, [90].

it may well be that our method scores better because it is faster to comprehend. If so, the difference in score would probably disappear or change if the subjects were given more time or improved training. In that case, we have not demonstrated increased effectiveness but that our method is easier to learn. This is a good thing, but there is a discrepancy between the real meaning of "more effective" in the hypothesis and our operationalization in the experiment. In other words, we do not have construct validity.

Some significant threats to construct validity are summarized below:

- *Definition:* Simply put, evaluating a hypothesis requires a precise understanding of what it claims. The language of science is not as straightforward as many like to believe. The same term or concept is often used differently, even in the same discipline. Hence, the conceptual apparatus that the evaluation is based on must be carefully defined.
- *Interference:* Different research programs, experiments, or measurements can affect each other. For example, suppose we aim to establish that specific changes in behavior patterns among school children are due to a new educational principle. In that case, the school children selected for our evaluation must not at the same time be exposed to other changes that may have the same or a partially overlapping effect.
- *Realization:* Any realization has inherent weaknesses – for example, when we operationalize concepts and definitions, apply methods and techniques, and make measurements. The realization must be robust to achieve construct validity. If we operationalize the same construct in a different but seemingly equivalent manner, the results should be almost identical.
- *Lack of width:* If we assert something, it must be studied or observed at full width.

  - Einstein's theory of gravity, for example, is not universal because it does not apply to subatomic particles (the micro-level). However, this cannot be observed at the macro-level.
  - A medicine has the desired effect only for specific dosages. The extent of this may not be detected if our observations are limited to only a few.

- *Side effects:* Solving a problem may generate new problems.

  - A new medicine may be effective for the disease it was designed to cure but have unexpected side effects.
  - A security upgrade of an IT system can remove some vulnerabilities while giving rise to others.

  An evaluation that does not uncover the side effects is not construct valid since the impact of the operationalization is only partly captured.
- *Hypothesis guessing:* Subjects are not passive participants. They may, in some cases, guess the researcher's hypothesis and change their behavior based on this. Some are uncomfortable with being observed. This may cause them to behave

differently and get a weaker score. Others become keener to appear in the best possible light. The scientists themselves can also easily influence the result, unconsciously and consciously. For example, whether or not the subjects are aware of the hypothesis in advance may significantly affect the outcome.

## 11.1.4 Conclusion validity

Conclusion validity concerns the relationship between the data collected and the conclusions drawn. Do we have sufficient evidence for our findings, or are they coincidental? More specifically, is our statistical analysis correct and the results significant?

**Definition 11.7** An *evaluation* is *conclusion valid* if the conclusions drawn are correct given the observations made.

*Example 11.4 (Lack of conclusion validity).* Again we reconsider Example 11.1. Lack of statistical significance is a common cause of inaccuracy, but we assumed this was not a problem in this example. Although the significance is acceptable, there may still be statistical reasons to doubt the validity of the conclusion. For instance, if the scoring metric is defined based on data already collected, the metric can consciously or unconsciously be fine-tuned to the advantage of our method. If it is conscious, it is called *(hypothesis) fishing*. Data can always be related in (infinitely) many ways, and some of these will, by pure accident hold, with statistical significance. Generally, it is good practice to define hypotheses, predictions, metrics, and scales before the evaluation is carried out and before initiating the data analysis.

Possible threats to conclusion validity include:

- *Needle in the haystack:* We may easily overlook essential aspects when the data volume is large and the data is heterogeneous. In that case, we can mistakenly conclude that present correlations do not exist.
- *Fishing:* As described in Example 11.4, we may always find significant relationships between data if we try sufficiently hard. Therefore, it is vital to start from already formulated hypotheses, predictions, and metrics and be true to these.
- *Assumptions:* Any statistical analysis is based on assumptions about data, methods of analysis, and the relationship between them. If the assumptions are not fulfilled, the conclusions can be wrong.

Figure 11.1 relates the four sub-concepts for validity. External and construct validity concern the generality. Can we defend our claims given the selection and grouping of subjects and the evaluation's setup? Are our concrete representations of concepts, constructs, and relationships satisfactory? Hence, external and construct

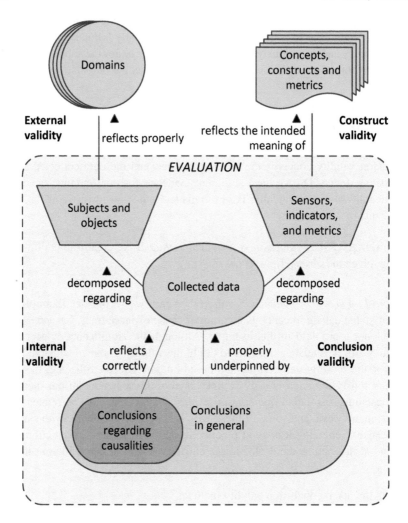

◀ reading direction

**Fig. 11.1** Relationship between validity concepts.

validity concern relevance for the external domain and the theoretical context. This is in contrast to internal and conclusion validity, which addresses the internal consistency. Are the conclusions correct for the data collected? Can we exclude alternative causes for the causal relationships we claim to have found?

## 11.2 Reliability

Reliability concerns the degree to which evaluations give the same results when repeated, partially or in full, potentially by others. It presupposes careful documentation of everything from setup to analysis of results. Reliability in the strict sense is unattainable for certain types of research, like action research.

Like validity, we decompose reliability into four sub-concepts: inter-observer reliability, internal consistency reliability, parallel-forms reliability, and test-retest reliability.

Most evaluations involve value-setting. This may be in the form of measurements, estimates, or judgments using quantitative or qualitative scales. The reliability of the evaluation depends on how these are defined and implemented. Some value settings depend on advanced instruments. Others are based on simple tests or questionnaires. If value settings are unreliable, most likely so is the evaluation as a whole.

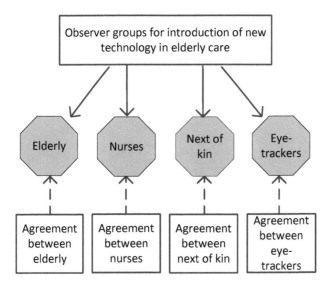

**Fig. 11.2** Inter-observer reliability.

### 11.2.1 Inter-observer reliability

Inter-observer reliability concerns the extent to which the observers' observations, assessments, and estimates correlate. The observers can be humans, but also animals, or technical devices. An evaluation may involve observers in different contexts. As Figure 11.2 illustrates, when evaluating a new technology for elderly care, we may need observer groups of elderly, nurses, and next of kin. In addition, the elderly can, for example, be equipped with eye trackers to document the degree to which they focus on the right things in the proper order. In other words, an evaluation may involve many observer groups. We have inter-observer reliability to the extent the observers agree between themselves in each observer group.

> **Definition 11.8** An *evaluation* is *inter-observer reliable* if the observations (assessments/estimations) in each observer group conform with each other.

*Example 11.5 (Lack of inter-observer reliability).* We again consider the security-testing experiment from Example 11.1. Since we use two groups of subjects, inter-observer reliability is relevant. If the scores in each group conform with each other, this is a positive indication of reliability. On the other hand, if this holds for only one group, while the subjects of the other group either did everything well or scored very poorly, there is cause for concern. It could, for example, indicate that there is a kind of threshold caused by a misinterpretation or an error in the setup.

### 11.2.2 Internal consistency reliability

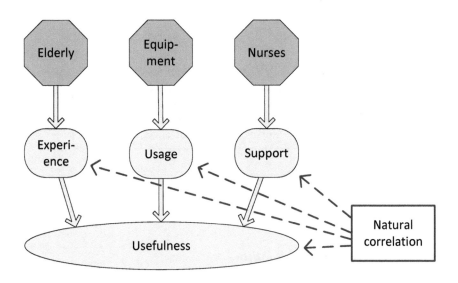

**Fig. 11.3** Internal consistency reliability.

Assessments, estimates, or measurements are commonly decomposed into several elements. Internal consistency reliability concerns whether there is a reasonable mutual correlation or correspondence between these elements and between each and the whole. For example, in Figure 11.3, we expect a positive correlation between user experience, correct usage, and good support. Furthermore, better user experience, usage, and support should increase usefulness.

> **Definition 11.9** An *evaluation* is *internal consistency reliable* if there is a natural mutual correlation between the results from different value-setting elements and between value-setting elements and the overall outcome of the evaluation.

*Example 11.6 (Lack of internal consistency reliability).* Assume that each subject in Example 11.1 completes a usability questionnaire and that we measure usability for each subject individually based on the completed form and the score in the experiment. We may have a problem if the correlation between the completed forms and scores is weak. Have we designed the form properly? Have the subjects taken filling in the form too lightly? Is the scoring metric used in the experiment sound?

### 11.2.3 Parallel-forms reliability

The same questionnaire can be expressed in a different but equivalent manner. We can change the order of questions, reformulate questions without changing the meaning, group them differently, and so on. Parallel-forms reliability implies we will get close to identical results independent of which equivalent alternative we use. Other evaluation methods can be varied accordingly. The distance traveled by a test subject can be measured with GPS, pedometer, or measuring wheel. Figure 11.4 extends Figure 11.3 with a variant or alternative to the original evaluation method to check parallel-forms reliability.

> **Definition 11.10** An *evaluation* is *parallel-forms reliable* if the choice between variants of the same evaluation method has almost no impact on the result.

*Example 11.7 (Lack of parallel-forms reliability).* Considering Example 11.1 once more, one approach to check parallel-forms reliability is to perform the same experiment on other web applications that, from a testing perspective, should be equivalent. Given that the subjects take both tasks equally seriously, a lack of conformity between the outcomes will cast doubt on the reliability of our initial findings.

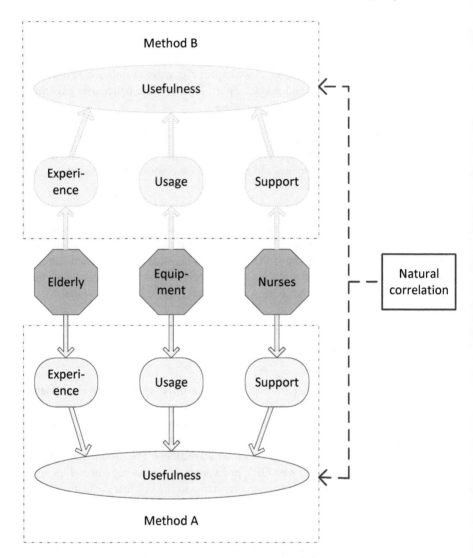

**Fig. 11.4** Parallel-forms reliability.

## 11.2.4 Test-retest reliability

When assessing repeatability, it is essential to distinguish time-independent parameters (constant over time) from parameters that vary over time (functions of time). For example, suppose we measure the volume of a human skull. In that case, we expect a reliable measurement to give almost the same result if repeated 100 years later (given that the human skull is not damaged in the meantime). In other words,

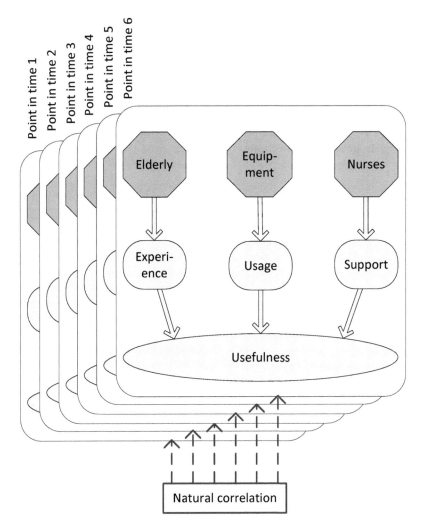

**Fig. 11.5** Test-retest reliability.

reliable measurement of a time-independent parameter may be repeated at a later date with almost identical results. As illustrated in Figure 11.5, test-retest reliability characterizes the degree to which the evaluation is time-independent.

**Definition 11.11** An *evaluation* is *test-retest reliable* if the result remains approximately the same independent of when it is repeated.

The size of the human population on Earth varies over time. If the hypothesis being evaluated characterizes the human population as a function of time, we may still have test-retest reliability. Otherwise, we may need another method to conduct the retesting. We may count from an airplane to estimate the number of bears in a specific mountain range this summer. To check the correctness of this counting in the future would require another approach. We could, for example, try to use satellite images from the relevant summer season.

*Example 11.8 (Lack of test-retest reliability).* We expect a high test-retest reliability for the experiment in Example 11.1. Low test-retest reliability is still possible if the number of subjects is low, if the subjects do not take the experiment seriously, or if the subjects are skewed concerning competence.

# Chapter 12
# Publishing

Publishing is an essential aspect of science. There are many different publishing channels. How to design, compose, and produce a publication depends on the content, publishing channel, and purpose. A publication is not necessarily written. Sometimes it is more appropriate to make a movie, a podcast, or give a presentation. In this book, however, we limit ourselves to written publications. Publications in technology science will usually devote considerable space to a description of the invention – the new artifact that the research has resulted in.

This chapter is divided into two parts. The first part gives an overview of available publication channels. The second part looks into ethical issues related to publishing, including the reuse of one's own already published material, also known as self-plagiarism. The topic of publication is followed up in Chapter 13, where we offer recommendations on the editing and structuring of technology science articles and advice on the writing process.

## 12.1 Selection of publication channel

Which publication channel we should choose depends on, among other things, the quality of the results, where we are in the research process, the kind of research we conduct, as well our own and any co-authors' expertise and experience. If we lack publishing experience, selecting a less demanding channel is usually wise. In the following, the various publication channels are ordered based on how difficult it is to achieve a satisfactory but not necessarily outstanding result. We start with the easiest, namely a scientific poster.

### 12.1.1 Scientific poster

A poster is especially suitable for publishing early ideas and slightly immature results. A poster is nothing more than a big sheet of printed paper. Most posters are presented orally by at least one of the authors, preferably by standing next to it and pointing or referring to its various elements. Therefore, the poster must be designed and organized to allow presentation in such a manner. We may, for example, let the poster tell a story or split the poster into components so that the intended message is conveyed step by step.

Typical beginner mistakes are using too much text and cumbersome academic sentences. Another common mistake is including irrelevant items, such as unnecessary logos and long-winded definitions. A good test of whether some element is relevant or not is whether it improves or simplifies an oral presentation of the poster.

If you are inexperienced, it is quickly done to design the poster so that the various elements point in different directions or diverge. You may easily end up with a poster whose ingredients lack a common thread. It is therefore advisable to start the poster design by identifying the three to four main points it should convey, such as the main conclusions, and then select or produce text, figures, and charts to support these points in a comprehensive manner.

### 12.1.2 Scientific abstract

A scientific abstract is a stand-alone document that summarizes some scientific work, typically of one to three pages. It is expected to have clear conclusions and contain a high-level argument for its correctness. Furthermore, it must describe the problem addressed in addition to relevant literature. To fit in everything we think "must be in" or identify what should be prioritized and what should be left out are well-known challenges for this kind of publication.

The term abstract is also used in other contexts or with different meanings. Some abstracts are longer and are often referred to as extended abstracts. They may be understood as an intermediary between a scientific abstract and a scientific article.

Most scientific articles also contain a so-called abstract. It is usually short, five to fifteen lines, and placed first in the article. It summarizes the article, in some cases including structural information. We will return to abstracts of this type in Chapter 13.

A so-called executive summary is also a kind of abstract. Technology science is often mission-oriented, that is, applied research conducted on behalf of a company, institution, or equivalent to solve a concrete problem. Then it is crucial to present the results in a clear and easily understandable form so that decision-makers can quickly understand the results and make decisions without reading the publication in full detail. A good executive summary is not more than one or two pages long and summarizes the most important findings in a structured and tidy manner. Structural information regarding the organization of the underlying report does not belong in an executive summary.

### 12.1.3 Popular scientific publication

A popular scientific article comprehensively presents research issues and findings to a broader audience. The theme may be limited to a particular field of interest and may or may not require some basic knowledge for the reader to benefit fully. For example, a zoologist may present new knowledge on the navigation techniques of migratory birds in a form that is easily understandable to hobby ornithologists; however, less suitable for the general public.

A popular scientific article usually presents new knowledge and only, to a small extent, the evidence or arguments for its correctness. In technology science, this means new artifacts and their potential utilization.

Writing a popular scientific article is a bit like writing for a newspaper. We must capture the reader's interest with a suitable headline, preamble, or picture and then present the material, so the reader does not lose interest.

A research-based feature story is a special kind of popular scientific publication allowing the researcher to actively participate in societal debate, using facts and knowledge from research.

### 12.1.4 Scientific article

There are various definitions of a scientific article. Our definition is as follows.

**Definition 12.1** A *scientific article* is an article that presents new, not previously published knowledge documented in a testable manner.

Some definitions require a scientific article also to be peer-reviewed and published in a journal. However, although assessment by competent peers may be of great help, it is not a feature of the article itself, and it is not difficult to find peer-reviewed articles of poor quality. Publication in a journal is also not a quality guarantee, as untrustworthy and completely frivolous actors flood the market. Furthermore, it may be more difficult in many disciplines of technology science to get an article accepted at a top conference than in many serious journals.

There are good and bad journals and good and bad conferences. It is not where the article is published or who has reviewed it that should decide whether an article is scientific or not, but its actual content.

The quality of a scientific article depends on many factors. The most important one is the new knowledge the article contributes. If the knowledge is not new but published, the article is not scientific in the strict sense. An article with groundbreaking new knowledge will typically be more highly rated than an article whose knowledge contribution is more average. But there are exceptions; for example, usefulness is also essential. An article with new knowledge of great relevance or potential for use is often ranked ahead of articles that contribute more knowledge but are of little use. How the new knowledge is presented and argued for is also essential. The history of technology science is full of examples where the same technology is invented more or less at the same time by independent research groups. In such cases, it may be the article that presents the technology in the most comprehensive manner that is cited and "remembered" in retrospect.

Scientific articles are published not only in journals but also in article collections or proceedings before or after scientific conferences, seminars, and workshops or in stand-alone books. Scientific articles have been, and still are, the "gold standard" of scientific publishing. This book has therefore dedicated Chapter 13 to article writing.

## 12.1.5  Scientific report

The requirements for a scientific report are essentially the same as for a scientific article. But scientific reports are not subject to the same expectations regarding brevity and readability imposed by commercial publishers. A scientific article is often a summary of one or more scientific reports. The validity and reliability of the scientific article may depend on documentation only found in the reports. It is then crucial that the article refers to these reports so that those who want to check or test the results know where to find the details. Suppose there is just a single report, and we want to avoid unnecessary editing and potential inconsistencies between the article and the report. In that case, a good solution might be to organize the report so that it consists of the published article followed by several attachments presenting the material and the documentation that could not fit into it.

In general, scientific reports should be structured in the same manner as scientific articles, a topic we will return to in Chapter 13.

Figure 12.1 relates scientific reports and articles to the publication forms described above. A scientific abstract can be seen as a specialization or refinement of a scientific poster (it may be the other way around if the abstract is short). Quite analogously, a scientific abstract can be detailed and extended into a scientific article. For all three, detailed documentation in one or several scientific reports may be required. Scientific articles can also be classified concerning publication channels. The requirements for a journal article are generally more stringent than those for a conference article or an article published as a book chapter, which are stricter than the requirements for a workshop article.

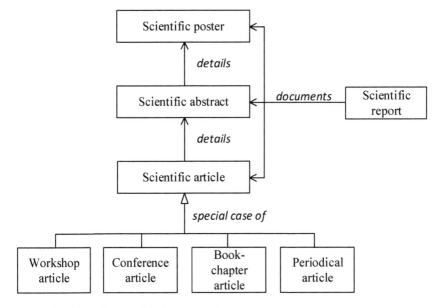

**Fig. 12.1** Publishing forms and their relationships.

## *12.1.6 Master's thesis*

A master's thesis is written to obtain a master's degree at a university or college. As a piece of writing, a master's thesis is a bit like a report, often of more than 100 pages. The extent to which a master's thesis can be called scientific varies greatly. Some master's theses are of high quality and give rise to one or more scientific articles. Others may be of high standard methodologically but do not contribute new knowledge. Others also have methodological weaknesses but are still sufficient for a master's degree.

A master's thesis could be structured following the recommendations for a scientific article described in Chapter 13, but there will not be the same constraints on length.

## *12.1.7 Doctoral thesis*

A doctoral thesis is written to obtain a doctorate at a university or college. A doctoral dissertation is expected to satisfy all requirements of scientific work.

In the past, a doctoral thesis was always a monograph, but today it is often a collection of scientific articles with an initial summary of up to 100 pages. The summary is expected to explain the relationship between the scientific articles and present the doctoral work as a coherent whole.

### 12.1.8 Scientific book

Books must also meet the same requirements as articles to be scientific. Of course, it is both possible and common to write books about science that do not present new knowledge, for example, textbooks, popular science books, or books that summarize a field of research. Scientific books are often significantly more educational than scientific articles, and the contribution to new knowledge is that they, in a new way, glue together fragmented knowledge into a whole. Writing books is demanding because there is much text and other material to keep track of. Moreover, the writing process itself is challenging. It is quickly done to change the style or level of abstraction along the way, without noticing it yourself, leading to considerable rework.

### 12.1.9 Patent

The preparation or writing of patents or patent applications is challenging for most researchers, not least because of the relevance of legal expertise. A patent may be defined as follows [97].

**Definition 12.2** A *patent* is an exclusive right to a product or a process that generally provides a new way of doing something or offers a new technical solution to a problem. To get a patent, technical information about the invention must be disclosed to the public in a patent application.

Getting a patent approved for some invention requires a level of precision that makes it possible to justify that the invention is new and not fully or partially covered by existing patents.

## 12.2  Reuse

It is our responsibility that our publications are up to the expected standard. This includes fairness and honesty in the planning and execution of the research process, artifact descriptions, and documentation of evaluations. It also means that we give others the credit they deserve and do not copy, reuse, or exploit our or others' work without this being agreed on and approved in advance by the publishers and the other relevant stakeholders involved. This may seem straightforward in theory but can be challenging in practice.

There is no doubt that today's pressure on publishing significantly impacts how scientists work, and not least publish. Nobel laureate Peter Higgs published fewer than ten articles after his famous paper on the Higgs particle in 1964 until his retirement in 1996. In an interview with The Guardian [1], he doubts he would obtain an academic position today because he would not be considered sufficiently productive. He also believes that under the current working conditions, he would not be given sufficient time and peace to repeat the work for which he received the Nobel Prize.

Today's researchers are forced to publish frequently. The number of publications is essential when applying for scientific positions and promotions, for example, from associate professor to professor or from researcher to senior researcher. The number of publications is also vital in the fierce fight for research funding. Furthermore, if you have been lucky and obtained funding for a research project, the granting authority will pressure you to publish widely and through various channels.

In the light of this, in recent years, there has been much focus on the reuse and exploitation of one's own work in new publications. The phenomenon is known as duplicate publication or self-plagiarism.

**Definition 12.3** *Duplicate publication* is the reuse of one's own already published material in such a way that this material in the new publication appears new and original.

Duplicate publication is problematic for publishers because they risk publishing articles that have long since been published or are being published by others. The consequence can be reduced news value, fewer citations, and ultimately decreased earnings. Duplicate publication is also a problem for conference organizers and initiators of seminars because they risk that some presentations have been given elsewhere in the past and therefore do not motivate people to sign up. Duplicate publication is also a problem for hiring committees and project evaluators because some candidates may appear better than they are since they have published the same work several times, under different titles.

Like for so much else, it can be difficult to distinguish right from wrong. For example, how much new material must an article contain to be published as new? What about a standard text, like the description of a research method the same authors have used and described several times in previous articles? Besides, how relevant is the issue of duplicate publication across different publishing channels? For example, are we required to refer to a popular scientific article if it appears before the scientific article?

All of this is disputed. We publish to spread knowledge that potentially may move humanity forwards. In such a context, it seems irrelevant whether authors reuse routine paragraphs they have written themselves, but there are different opinions here.

Deliberately misinforming the outside world by presenting old results as new is reprehensible. On the other hand, it is crucial not to lose sight of the purpose and importance of dissemination. There is a reasonably broad consensus that the following is unproblematic:

- for a scientific article, there is also a scientific report that either is identical to or an expanded version of the article;
- publishing preprints via arXiv.org – in fact, almost everything published in specific fields of physics and mathematics is made accessible through this portal;
- to send an article published in a conference proceeding to a journal for review as long as we simultaneously inform the journal editor and we do not break agreements related to its already published version;
- to accept an invitation from a serious journal to republish an article as one of several selected articles after a conference;
- to "follow the path" from the publication of abstract or poster via publication of workshop or conference article to a journal article, with the same piece of research, but in an increasingly more thorough and complete form. Again it is essential to play with open cards and carefully specify the contribution of the latest version compared to what you have already published.

# Chapter 13
# Article Writing

The scientific article is the gold standard of scientific publishing. This is indisputable for basic research. To a great extent, it also applies to technology science. Therefore, it makes good sense to emphasize article writing in a book like this.

## 13.1 Structure

Scientific articles can be structured in many ways, with few absolute rules. However, a typical technology science article consists of six main parts, as illustrated in Figure 13.1. The introductory part includes the title, author list, abstract, keywords, introduction, and characterization of artifact needs. The extent to which the artifact needs are presented in a separate section or integrated into the introduction depends on the artifact in question – not least the complexity of the needs. The introductory section may also present background material, for example, existing relevant technology, which is ignored in the following, as background material usually must be sacrificed on the altar of space restrictions.

The research method part is located immediately after the introductory part. It may be decomposed into sub-parts addressing the needs, the invention, and the evaluation phases. Alternatively, the research method for the evaluation may be presented in the evaluation part of the article. The sub-parts may, of course, be further decomposed as required.

The purpose of the artifact part is to describe the invention. The exact approach depends on the type of artifact in question, but some kind of artifact description is necessary. The description will typically characterize the artifact's architecture, its behavior, and the characteristics of essential components.

The evaluation part presents results and experiences from investigations, studies, or experiments conducted to evaluate and test the features and qualities of the invention. Its structure depends on the discipline, artifact type, and selected methodology.

© The Author(s), under exclusive license to Springer Nature Switzerland AG 2023
K. Stølen, *Technology Research Explained*, https://doi.org/10.1007/978-3-031-25817-6_13

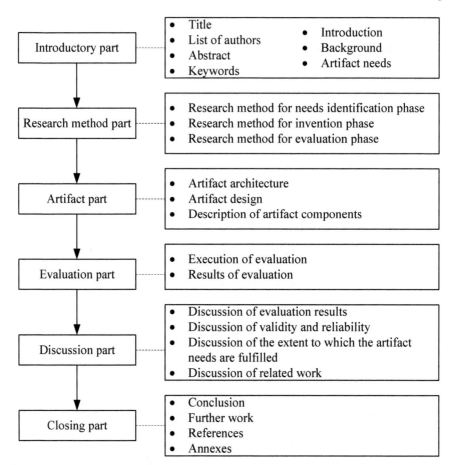

**Fig. 13.1** Breakdown of a technology science article.

It is expected to document the evaluation and results in a testable manner, including the collected data and measurements and their analyses.

In science, few claims can be made with absolute certainty. A discussion part is therefore also needed. Relevant discussion issues are the evaluation results and their validity and reliability. The discussion should also cover the degree to which the artifact needs have been fulfilled and the relationship to other work and existing literature.

The closing part consists of a conclusion, plans for further work, thanks to partners and contributors, as well as references. It may also contain annexes, given that there is room and need.

As pointed out in the previous chapter, a scientific report is, in many ways, a scientific article without space restrictions. The recommendations given below are therefore also relevant for report writing. The same applies to master's theses and doctoral theses written as monographs.

In the following, we address these main parts in more detail, emphasizing the specific challenges of technology science.

## 13.2 Introductory part

The introductory part serves several purposes, as already explained. It should advertise the work to potential readers and provide the context and setup for the rest of the article.

### 13.2.1 Title

The title of the article is vital for several reasons. First, it should, in a compact manner, tell potential readers what the article is about and why they should read it. Second, the phrases and terms it contains influence under what conditions and to what extent search tools retrieve the article.

The title is also essential for the writing process because it should carefully capture the main contribution or topic of the article. There should be a common thread from the title via the various structural elements to the main conclusions.

The title must not be too general. It should convey the essence of the article's contribution. On the other hand, it must not be too eccentric, using terms, concepts, and acronyms that relevant readers do not understand.

A descriptive title focusing on the new artifact and its advantages is often a good choice. If, for example, the topic is robot-based combating of brown snails, the following titles are of this kind:

- Design of a robust robot for picking brown snails in the garden environment.
- Algorithm for identifying brown snails in garden-environment images.
- Grab arm for picking brown snails from the lawn and in flower beds.

A title in the form of an assertion may also work well. The assertion could be the central hypothesis or the conclusion. Examples in this direction are:

- Reliable and safe robots for picking brown snails can be produced at consumer-friendly prices.
- Algorithm for identifying brown snails in garden images with an error rate of max 0.1%.

Titles in the form of puns or literary quotes, often used in newspapers and magazines, are usually not recommendable, at least not standing alone. They are often overlooked or misclassified by search tools.

It is recommendable to analyze potential titles term for term. Concepts and phrases that hardly contribute or are not meaningful should be removed. Testing the title on colleagues, family, friends, and similar is often helpful. Understanding much of the content is not necessarily required to contribute constructive suggestions for how the title should be improved.

### 13.2.2 Author list

The author list provides the names of the authors. The author list is often the subject of discussion or, at worst, conflict. The disagreements revolve around two issues:

- *Authorship:* Who has contributed sufficiently to be included in the author list?
- *Order:* Which naming order convention should be used, and if so, how should it be implemented in the present case?

Concerning *authorship*, there are recognized conventions and recommendations to lean on. According to the Vancouver rules [40], which target medicine but have far broader relevance, every author should have participated in the work to such an extent that the author can take public responsibility for the relevant sections of the content. One or several authors must furthermore be responsible for the integrity of the work as a whole, from planning to publication. Authorship should solely be based on:

1. substantial contributions to the conception or design of the work; or the acquisition, analysis, or interpretation of data for the work; and
2. drafting the work or revising it critically for important intellectual content; and
3. final approval of the version to be published; and
4. agreement to be accountable for all aspects of the work in ensuring that questions related to the accuracy or integrity of any part of the work are appropriately investigated and resolved.

These rules are also subject to interpretation. If we still doubt whether a contributor should be included in the author list or not, it is usually a good idea to give the benefit of the doubt to the person in question.

In terms of *order*, the two most common conventions are:

- Alphabetical ordering according to the family name.
- Ordering according to the size of the contribution.

The first one is easy to implement but is best suited when the contributions of the authors are close to equal. The other is fairer but is more challenging to implement because measuring one contribution against another can be problematic. In some research groups, the first author is the one that does most of the writing, but this works poorly if others made the main scientific contribution. For an article written by a doctoral or master's student, it is common practice that the student's name comes first, followed by the supervisors.

Discussing and agreeing on these issues in advance is wise to avoid ordering conflicts. Who is responsible for what, and in what order should the authors be named? You may also agree on a procedure to resolve potential disputes. Then everyone knows the rules of the game and can plan their effort and time accordingly.

### 13.2.3 Abstract

Most scientific articles contain a short section referred to as an abstract. Usually, this abstract is placed immediately after the article title and the author list but before the introduction. An abstract in a scientific article is not a "scientific abstract" in the meaning of Section 12.1.2, but a summary of five to fifteen sentences – in some cases, including structural information.

A good abstract reflects both the title and the article's main contribution. Unfortunately, it is not uncommon to come across articles where the title and abstract point in different directions.

The abstract plays, to some extent, the same role as the article's title. If a potential reader finds the title attractive, the abstract is usually the next thing they look at. It is therefore essential to formulate the abstract so that intended readers do not lose interest. Many search tools also process the abstract to select which articles to display. We should, therefore, carefully choose concepts and formulations to ensure that the article is picked up when relevant.

### 13.2.4 Keywords

The keywords classify articles concerning content. Carefully selected, they allow their intended readers to quickly find the article when searching. It is usually a matter of identifying five to ten keywords. Some periodicals operate with lists of predefined keywords from which the authors can choose.

### 13.2.5 Introduction

Among other things, the introduction has a motivating role. It should convey why the article is interesting and relevant to its intended readers. The introduction should be comprehensible but at the same time concise so that it is not perceived as boring.

The introduction should clarify the artifact needs and how and to what extent they are addressed. It should also provide some overview of the field and its research front and, based on that, characterize how the article improves on the state of the art.

As explained in Section 6.1, it is often the case that hypotheses are not expressed explicitly. This is unproblematic as long as they implicitly follow from other documentation in the article. On the other hand, using a hypothesis that is not overly technical, such as a working hypothesis on which we based our research, to build or structure the introduction may improve its readability and motivate the reader.

The introduction should provide an overview of the various steps and activities we went through to arrive at our results. It may be beneficial to base this overview on a figure. In a scientific article, there is no reason to keep the main findings secret from the reader until the end. What we see as the article's actual value, and its contributions of which we are proud, should be highlighted in the introduction to sharpen the reader's appetite.

Readers of scientific articles like to know the "path they are to follow" before they start walking. Hence, the introduction should prepare the reader for the rest of the article. This involves presenting and motivating its breakdown and giving an overview of what is described where. If the article has attachments, their content should also be outlined. It is annoying to discover an attachment with definitions after first having read the core part of the article. If there are scientific reports or permanent websites with background material, these should also be mentioned.

### 13.2.6 Characterizing artifact needs

Technology science is about fulfilling (possibly potential) artifact needs. The article must present these needs. The level of detail depends on the available space and the kind of artifact in question. If the needs and the uniqueness of our contribution are easy to describe, a paragraph or two in the introduction may be sufficient. In other cases, a separate section may be required. Whether we describe the needs in prose, as requirements, or as success criteria is insignificant as long as they are communicated clearly at a suitable level of abstraction. Due to space restrictions, it may be necessary to simplify or summarize the needs at a high level of abstraction. If so, we should make the detailed needs available via other means. A technical report with full details is often a good choice.

## 13.3 Research method part

The research method is the specific approach or procedure according to which we perform the research. In technology science, this means how we go about inventing an artifact and evaluating to what extent it fulfills the identified needs. The artifact

itself may be a new method – for example, a new and better method for early identification of cyber-attacks. This new method must not be confused with the research
method used to invent and evaluate this new method.

The description of the research method is an often-neglected part of technology
science publications. One reason could be that the literature on research methodology mainly addresses natural and social sciences and uses terminology that seems
foreign to many technology scientists.

It is often helpful to divide the research method part into three sub-parts:

1. *Research method for the needs identification phase:* What we did to capture the
   artifact needs.
2. *Research method for the invention phase:* What we did to invent the new artifact.
3. *Research method for the evaluation phase:* What we did to evaluate the new
   artifact.

These can then be further decomposed as needed. The following must be described:

- *materials* – required in prototypes or to try out and test the artifact;
- *procedures* – including technical approaches, standards, and so on;
- *equipment* – including setup, parameters, and the like;
- *analysis* – including statistical analysis, performance analysis, economic analysis, and the like.

The identification and analysis of needs may require special studies. It may also
be necessary to perform dedicated literature surveys or investigate the properties of
different materials relevant to the artifact. For each such study of some significance,
the research method should be described or at least documented by a reference to
such a description. For idea generation, documenting the process itself is not always
important. If we get a brilliant idea while having breakfast, this is irrelevant to the
article, even though the whole article is based on this idea. On the other hand, in the
case of action research, if we follow a detailed procedure for action identification,
this procedure must be documented.

## 13.4 Artifact part

How the artifact is described depends, of course, on its kind. In extreme cases, the
artifact description is just half a page; in others, it is the largest part of the article.
Often the artifact is described from different points of view and at varying levels
of abstraction. Usually, there is at least some kind of design specification. It may
be an architectural drawing if the artifact falls within building and construction, a
chemical formula if the artifact is a new material, or a flow chart if it is a new
work process. A design specification describes the main components of the artifact,
how these components influence, communicate, and interact with each other, and

what each component does or offers. The components may be substances, materials, machines, software, human processes, activities, or structures.

## 13.5 Evaluation part

The evaluation is about testing, examining, and checking whether or to what extent the invention meets the artifact needs. As described in Chapters 7–10, how this is done depends on the research field and the kind of artifact in question. In many cases, however, some form of prototyping is required. Documenting the prototype and to what extent it mirrors the proposed design is vital for the validity of the results.

Often, we must combine several evaluation methods. In that case, the evaluation part must explain how they are integrated or cover different aspects or qualities. It may often be helpful to structure the evaluation part to reflect this.

A key ingredient is the presentation of evaluation results, including statistical analysis to the extent this is relevant. Consistency, tidiness, and a suitable structure reduce the likelihood of misinterpretations. In the case of insufficient space, there should be links or references to the missing documentation (for example, to an open scientific report).

## 13.6 Discussion part

An essential part of a scientific article is the discussion. Scientific answers and conclusions are mostly not absolute. There is almost always some room for doubt. For example, the conclusion depends on or involves all sorts of assumptions, abstractions, and uncertainties. In the following, we have a closer look at some issues that the discussion part should cover.

### 13.6.1 Discussion of evaluation results

Mixing the presentation of the (more) objective results like measurements, filled-in questionnaires, various forms of analysis, and so on with their subjective interpretation is a common mistake in article writing. The reader may easily misinterpret personal views and impressions as more "objective" than they are. We recommend separating the evaluation results and their discussion structurally. The reader then knows what the authors claim to belong to which. It may also simplify the writing process since it helps the authors to keep the two things apart.

Suppose the evaluation is composite, consisting of several sub-evaluations (different studies, experiments, surveys, etc.). In that case, there might be some significant textual distance between the presentation of the results of one sub-evaluation and the discussion of these results. The left-hand side of Figure 13.2 illustrates the problem. The discussion comes after the results of each sub-evaluation have been presented. If so, restructuring according to the right-hand side of Figure 13.2, where each sub-evaluation has its section structurally decomposed into a result and a discussion part, might be a good idea.

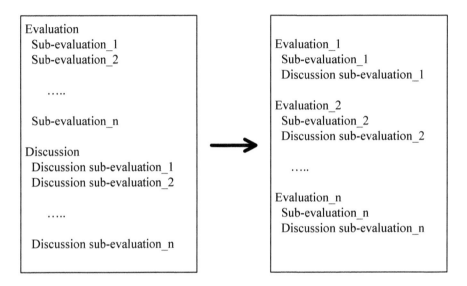

**Fig. 13.2**  Structure for composite evaluations.

## 13.6.2 Discussion of validity and reliability

Empirical evaluation is challenging, and there are many possibilities for mistakes. The discussion part should, therefore, carefully assess threats to validity and reliability as described in Chapter 11. This applies to the evaluation as a whole and each sub-evaluation. This kind of discussion is often demanding from the authors' perspective. It may feel a bit like sweeping away the foundation for your research. This is especially true if the object of evaluation includes human activity or behavior, due to the many possibilities for unwanted interference and side effects.

Either way, this is about being honest and not pretending that your findings are better than they are. It may be helpful to structure the discussion according to the breakdown of validity and reliability in sub-concepts (see Chapter 11).

### *13.6.3 Discussion of whether artifact needs are satisfied*

Even after extensive evaluation, it is not necessarily apparent to what extent the artifact needs are satisfied. It is not uncommon that the artifact needs are only fulfilled partly – this does not necessarily reduce the quality of the work or make an article less interesting. We may have succeeded with some of the needs, less well with some of the others, while it is unclear to what extent we have succeeded with the remainder. In that case, it might be a good idea to address one need after another and thereby provide the reader with a neat and understandable picture of the situation.

### *13.6.4 Discussion of related work*

There are various views and traditions concerning whether literature and related work should be discussed in the introductory part or towards the end of the article. There is no clear-cut answer. It depends on the subject field and the complexity of the artifact. In general, in technology science, detailed literature discussions usually fit better late in the article because the reader may need insights into the artifact constituting the main contribution to appreciate the discussion fully. This does not mean that the introduction or other parts of the article should refrain from discussing related work, but a detailed literature discussion usually becomes better if we can refer to aspects of the invented artifact, which is difficult to do before it has been presented.

## 13.7 Closing part

Inexperienced authors put too little emphasis on the closing part. They tend to think that this part may be compiled easily or with little effort because the real contributions have already been presented, and the reader can deduce the main conclusions. This is a mistake.

### *13.7.1 Conclusion*

The whole article should support its conclusion. Sections that fail to do so can, strictly speaking, be left out. Unfortunately, some authors do not sufficiently prioritize the conclusion. Articles with no clear conclusion are pretty common. The results may still be of high quality, but this follows only implicitly from the scientifically challenging parts of the article. Thereby, the actual value of the contribution may not be understood.

It can be helpful to think of the conclusion as a stand-alone summary, somewhat in the spirit of the executive summary described in Section 12.1.2. What we want the reader to understand should be expressed explicitly and served piece by piece "with a teaspoon." The conclusion can be tedious and demanding to write, but it plays a crucial role in the process of result dissemination.

### 13.7.2 Further work

Commonly, scientific articles contain a section or paragraph describing ongoing work or further plans for the research in question. In general, this description should be brief. A description of further work is beneficial but not necessary, and it does not significantly contribute to the article's value.

### 13.7.3 Thanks

It is good practice for the authors to thank other contributors. This may be:

- assistants and partners who have helped with the research without qualifying as joint authors;
- colleagues who have read and commented on or contributed in other professional ways;
- financiers or equipment suppliers;
- employers, managers, or other facilitators.

Readers may implicitly assume named contributors to favor the positions taken in the article. If it is unclear whether someone wants to be named, we should obtain permission, preferably in writing.

### 13.7.4 Bibliography

If we make use of or build upon the work of others, this should be acknowledged in the form of a citation. However, this does not apply to work generally accepted or described in detail in textbooks. When citing, it is good practice to refer to the original publication. In addition, it is often required to include later works to the extent these contribute with better explanations or, for us, more relevant further developments.

Web references should also be included in the bibliography. If possible, enter permanent URL links. For all web references, the access date must be specified. It is good practice to download the cited content and store it securely in your archive since websites change or may be removed.

Most periodicals and conferences expect the references to be in a specific form, for example, under the Harvard rules [2] or the Vancouver rules [40]. These conventions are motivated by several considerations. They ensure that the cited contributions are more easily retrievable and classified concerning the type of publication, evaluation process, and the quality control it has undergone.

Unfortunately, there is much sloppiness and inaccuracy in implementing such conventions. The same publication type is frequently classified differently in the same literature list. A conference article may be presented as a book chapter at some point. In contrast, another conference article may be classified as what it is or as a periodical article somewhere else in the list. Alternatively, while one conference article may be specified in great detail, including the conference location and postal address, other conference articles may be under-specified.

A common cause of inconsistency, regardless of whether we use Endnote, Bibtex, or any other citation tool, is uncritical copying of references from various databases. To avoid this, experienced authors maintain a dedicated literature database based on a well-thought-out policy making sure that citation data from external databases are adjusted to comply with the policy as soon as they are downloaded.

### 13.7.5 Attachments

Attachments may be used to provide additional details, for example, to allow peers to check the validity of our findings. It is crucial to make the reader aware of the attachments already in the introduction.

## 13.8 If you get stuck

Writing a nontrivial piece of text can be challenging. Our ideas may feel immature, or we may struggle with the structuring, the sentences, etc. External factors may also complicate the writing process. There is no universal medicine. In some cases, however, there are a few things we may try. In the following, we present some of these.

### 13.8.1 Getting started

What should we do first, and how should it be approached? Our ambitions may be an obstacle. When we finally carve out a few sentences, they may seem far below par compared to our expectations. This may put us off and cause the whole enterprise to be abandoned.

In such situations, it may help to temporally put all ambitions on ice and specify a set of simple writing tasks that must be accomplished at some point to complete the article. Each task should have a clear goal and, when put together, make up a good starting point for the next iteration, which can be approached similarly. This way, we execute the writing process in a stepwise fashion. We may, for example, try to complete one new task per day or finish the first before the morning coffee, the second before lunch, and so on.

One possible task is to identify a working title, a second to sketch the abstract, a third to outline the table of contents, a fourth to identify a kind of structure for the introduction, and so on. By continuing in this manner, we obtain a first draft of the entire article step by step. Having reached this point, the writing process may feel easier because we have a draft to read and improve. If not, we can continue in the same style by identifying a new set of simple tasks. The proverb "many small streams make large rivers" denotes the process of becoming rich by accumulating wealth from many sources or small businesses; it may, as indicated above, also be seen as an idiom for article writing.

## 13.8.2  Establishing a common thread

In a well-written article, there is a common thread, from the title, via the abstract and the various sections constituting the body, and right up to the conclusion. An article in technology science should present the artifact needs it is supposed to fulfill, the artifact's main features, how it improves on the state of the art, and conclude on its suitability. Anything not supporting the conclusion is strictly speaking irrelevant. Hence, to help organize and plan the writing process, it can be fruitful first to outline the conclusion and then use this to guide the structuring of the article. We can use the same approach to determine the parts of a sprawling article that should be cut.

A frequent cause of unclear direction is extensive reuse of text. This may be text from the project application, the problem analysis, or introductory studies that we have completed. Reuse can be both wise and good, but only if the text meets the needs of the article. It may be demotivating if the text we have spent much time and diligence to produce at an early stage does not fit in, but we have to accept this. Most authors find that much of what they write in the end has to be scrapped or replaced by something better. The work required to produce the text no longer needed has usually nevertheless been valuable. It may have helped us further or given us new or improved insights.

If the co-authors' contributions are not adequately integrated, they will pull in different directions. A common cause is that no author takes the overall responsibility. Moreover, some are mainly interested in putting forward their own work or ideas. How to handle this depends on how well we know our co-authors, their flexibility, and our willingness to compromise. If there is consensus on the conclusion, we may use this to tighten and improve the presentation as suggested above. In the opposite case, splitting the article into two or more articles may be better, each addressing different aspects of the research.

The human tendency to prioritize the simple parts is another possible culprit. In article writing, this means writing text we feel comfortable writing. Typically, the result is large background sections presenting essentially irrelevant material but easy to write because we can build on what others have written. In such cases, more discipline is required. If there are sections that seem particularly challenging, the strategy of decomposing into simpler sub-tasks that can be addressed one by one, as recommended in Section 13.8.1, is a way forward.

Publishers of scientific articles tend to disallow tables of contents, which may be unproblematic for the readers since most articles are short. A table of contents is nevertheless a helpful tool to establish a common thread during the writing process. In a well-written article with suitable headings, the contents list can be read almost like an abstract. To evaluate or check the suitability of the article structure, try reading the automatically generated table of contents as a piece of prose. To the extent there are giant leaps or apparent deviations in direction, adjust the headings until the problem disappears. If it does not, the article may need restructuring.

### 13.8.3  Little to discuss

If there is "little to discuss," the problem is most likely the introduction. If the needs are vague, fluffy, or absent, there is no firm ground for the discussion. In that case, we must characterize more precisely what the aims are. When the needs are appropriately specified, for example, as a bullet list, we can address them individually and discuss to what extent they have been fulfilled. The base for this discussion is the evaluation. To the extent the findings were surprising, conflicting, or need interpretation, this should be addressed. Their validity and reliability are essential issues in this context.

Lack of precision is also problematic for the literature discussion. A comparison with competing approaches will hardly be convincing unless our artifact is appropriately specified. If the literature discussion is brief because there is little to find, this should be documented. We may, for example, refer to an appendix or a technical report specifying the details (search tools, search terms, date of search, and so on).

If we have read or understood too little of the relevant literature, limited time is perhaps the indirect cause. In that case, we must streamline the reading. Inexperienced researchers tend to read articles, reports, and the like, too much from beginning to end. This is not surprising since this is usually what the authors of these publications had in mind when they wrote them. If we have gained experience or some expertise in a field, it may be just as effective and much more efficient to read backwards from the end or look up the main findings and read just enough to understand what they are. If there are certain aspects of these we do not understand or need to check more carefully, we search for the relevant explanations and read those. That way, we may save time and discover weaknesses we would not have found with a more conventional reading approach.

### 13.8.4 Conclusion says nothing

If the conclusion seems pointless and impossible to improve, maybe we have no results worth publishing. However, in many cases, the cause of the problem is located in the introductory part of the article. The conclusions tend to become likewise if the introduction is vague or imprecise concerning goals, aims, and direction. In that case, revisit the introduction and clarify what the new technology should be be good at. It could mean that we must also improve the evaluation part and, in some cases, also perform additional evaluations.

A weak conclusion may also be due to modesty or fear of being attacked. When we have selected our research method, carried it out to the best of our abilities, and argued for the validity of our findings, it is our duty as researchers to put them forward clearly and explicitly. It should be highlighted if we have arrived at results that conflict with or are better than others. If we have identified mistakes in the work of other researchers, we should say so explicitly.

### 13.8.5 Nothing more to cut

The publisher may impose strict constraints on the length of an article, and most authors will sooner or later experience having to cut into something that they think is virtually impossible to shorten. As already argued, it may be helpful to prioritize the various sections and paragraphs based on the degree they support the conclusion. Those that do not significantly impact the conclusion's validity are prime candidates for deletion or shortening.

Background material and details for which there is no room in the published article, but which are strictly required for others to control and test our findings, should be made available in some other form, for example, in a technical report or some look-up facility by the publisher. This may also be a safe haven for the material we are forced to leave out as part of the cutting process.

If the constraint on article length is expressed in the number of pages and not in the number of words, we may try to redesign figures and charts. In that case, it is essential to utilize space without compromising comprehensibility and legibility.

Circumventing length restrictions by ignoring or redefining the publisher's pre-defined formats is not good. First, this is usually discovered, with additional work for the authors and delayed publishing as the likely result. Second, if not detected by the publisher, it will, in many cases, reduce comprehensibility and harm the reputation of the article and its authors.

### 13.8.6 Artifact needs do not fit in

If the artifact needs are very extensive, they may take up too much space. In that case, they must be simplified or made more abstract. The detailed needs may be included in a supporting technical report.

The needs may have changed during the research process after they were characterized. The article may also address only some aspects of the needs. If so, they must be revised and updated.

It is possible that we initially had no clear idea about the needs and the new artifact arose more or less by chance, for example, when experimenting with new or exciting technology. In that case, the needs (possibly potential or future) must be identified as part of article writing.

There are few absolute rules for article writing. Still, in technology science, the introductory part should always present the artifact needs at a suitable level of abstraction – at least in a couple of sentences. They may be integrated into the introduction or assigned a separate section in the introductory part, as illustrated in Figure 13.1.

### 13.8.7 Hypothesis does not fit or is missing

As explained earlier (see Section 6.1), in technology science, it is often the case that the hypothesis implicitly follows from other documentation in the article. If we have characterized the artifact needs and described the artifact, there is an implicit hypothesis that an artifact built according to the description will fulfill the needs. We do not have to state the hypothesis explicitly as long as it follows from other documentation in the article.

However, incorporating a hypothesis in the introduction may be beneficial from a presentation perspective. This may be an early working hypothesis, for example, that a particular technology our artifact builds on is suitable to meet the identified need. While the detailed hypothesis occurs only implicitly, an explicit high-level working hypothesis in the introduction is often a good solution presentation-wise. The implicit hypothesis should then follow as a detailing or refinement of the working hypothesis.

### 13.8.8 Are predictions needed?

According to Definitions 7.1 and 2.13, an educated prediction is a hypothesis. As for hypotheses in general, predictions may be expressed implicitly if they follow from other documentation, but readability may be improved if they occur explicitly. It is not uncommon for authors to refer to predictions as hypotheses or sub-hypotheses, which is usually unproblematic. The terms and concepts we use are less important if we do the right things. Which wordings are best suited also depends on the research field and the kind of readers we are addressing.

# Chapter 14
# Technology Science from the Perspective of Philosophy of Science

This book is not about philosophy. Nevertheless, it is interesting and valuable to look at technology science from the perspective of philosophy, particularly the philosophy of science. This chapter aims to do just that. Section 14.1 gives an overview of the main directions of the philosophy of science, emphasizing the last 100 years. The presentation is simplified. We concentrate on the aspects of direct relevance for technology science. The presentation is also polarized. We tend to emphasize the more extreme views to highlight the differences between the various directions.

Section 14.2 positions technology science as defined by this book in the overview presented in Section 14.1.

## 14.1 Main directions of the philosophy of science

This section is inspired by several sources, but mainly by the book *What Is This Thing Called Science* authored by Alan Chalmers (b. 1939) [14].

### 14.1.1 Empiricism

The terms *a priori* (from the former) and *a posteriori* (from the latter) are used in philosophy to classify knowledge as to whether or not it depends on experience.

**Definition 14.1** *A priori knowledge* is knowledge that is not based on experience. *A posteriori knowledge* is knowledge that is based on experience.

Strict empiricism means that knowledge is obtained only a posteriori – in other words, experience is the only source of knowledge.

Modern empiricism has its origins in what is known as British empiricism. John Locke (1632–1704), often regarded as the originator of British empiricism, argued

K. Stølen, *Technology Research Explained*, https://doi.org/10.1007/978-3-031-25817-6_14

that all knowledge is a posteriori. We have no innate knowledge. At birth, human consciousness is like a blank sheet of paper, and this paper can only be filled with experience. All knowledge and ideas originate from experience. Experience is of two kinds, namely sensation and reflection. Sensation tells us about things and processes in the external world, while reflection makes us conscious of our mental processes [85].

### 14.1.2 Inductionism

Inductionism is a variant of empiricism dating back to Francis Bacon (1561–1626) [45]. Bacon, an important inspiration for British empiricism, describes the scientific method as a procedure that systematically gathers information and transforms this information into understanding. This procedure Bacon refers to as induction. The systematic collection of information results in a multitude of individual observations, for example, *n* observations of a particular experiment giving the same effect. The resulting understanding is a statement of validity beyond the individual observations, for instance, that the experiment always gives the same result.

Another prominent inductionist was John Stuart Mill (1806–1873) [95]. According to Mill, induction is a prerequisite for all meaningful knowledge, including mathematics. Mill argued that all mathematical truths, including rules of logical deduction, are generalizations from experience and based on induction. That we imagine something as logical or necessary is a psychological mechanism of man. We are unable to see other possibilities than those that logic and mathematics allow. Mill thus represented an empiricism whose extremism is difficult to surpass [34].

### 14.1.3 Positivism

Auguste Comte (1798–1857) is regarded as the originator of positivism [8]. Positivism argues that all real knowledge is scientific knowledge and that such knowledge can be obtained only using the scientific method. Positivism implies a natural scientific attitude to the study of man, as opposed to a normative perspective prescribing how man should behave.

Comte postulated that the evolution of humanity goes through three phases, namely the theological, the metaphysical, and the positive. In the positive phase, theology and metaphysics have been replaced by a hierarchy of sciences with sociology at the top. Mathematics is the foundation of all other sciences and is therefore at the bottom. As illustrated in Figure 14.1, Comte distinguishes between six sciences and classifies them in terms of complexity and positivity (positivity in the meaning of exactness). Complexity is inversely proportional to positivity. Mathematics is the most positive, while sociology is the most complex. A science builds

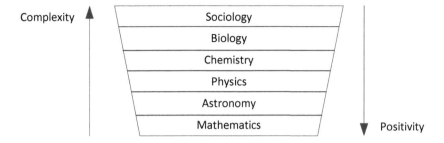

**Fig. 14.1** Comte's hierarchy of sciences.

on sciences with higher positivity and provides the foundation for sciences with lower positivity. For example, chemistry is based on physics, while chemistry is a foundation for biology.

### 14.1.4 Logical empiricism

Logical empiricism[1] denotes a philosophical direction but also a philosophical movement [16]. Logical empiricism flourished in the 1920s and 1930s, with Vienna and Berlin as geographical centers of gravity.

Logical empiricism combines strict empiricism with formal logic. The formalization of logic and mathematics in the late nineteenth century and early twentieth century was an important source of inspiration. In logical empiricism, formal logic and mathematics are instruments that separate the scientific from the not scientific. By combining Gottlob Frege's (1848–1925) thesis that all mathematical truths are logical with the early Ludwig Wittgenstein's (1889–1951) idea that all logical truths are logical tautologies, they concluded that all statements are either analytic (a priori) or synthetic (a posteriori). On this basis, they distinguished between hypotheses that have meaning and those that have none. This so-called demarcation principle postulates that a statement is meaningful only if empirically testable. In other words, any hypothesis that is not purely logical or cannot be tested empirically is meaningless. As a result, most metaphysical, ethical, aesthetic, and other traditional philosophical problems were reduced to pseudo-problems.

That something is empirically testable was interpreted differently in the different directions of the movement. The most extreme variant required verification in the sense of a definitive procedure for determining whether the statement is true or not. The demarcation principle then becomes a verification principle.

---

[1] Logical empiricism is also known as logical positivism. The former term is, however, a better fit given the definitions of empiricism and positivism in Sections 14.1.1 and 14.1.3.

## 14.1.5 Falsificationism

Falsificationism is a scientific direction that sharply distances itself from any verification requirement. According to a falsificationist view, a scientific theory can never be verified – it can only be falsified. Karl Popper is considered the originator of falsificationism.

The requirement for a definitive procedure to determine whether a hypothesis is true or not, as put forward by some logical empiricists, is very strict. For example, it implies that no universal hypothesis addressing a domain of infinitely many phenomena is meaningful. In other words, the following hypothesis, according to legend formulated and tested[2] by Galileo Galilei from the leaning tower in Pisa, is unscientific:

> If two cannonballs of different mass are dropped from the top of the leaning tower in Pisa, the time they spend until they reach the ground will be identical.

Although the same experiment is repeated thousands of times, and we get the same result each time, we can never be sure that the hypothesis is true. In principle, it could be that we only get the predicted result if the cannonballs are dropped at certain specific points in time, and since there are infinitely many of those, it is impossible to test all.

The logical empiricists were, of course, aware of the problems related to universal hypotheses, and several weaker variants of the demarcation principle were formulated. Popper, however, went furthest and introduced the falsification principle [63].

According to Popper, a hypothesis is meaningful if there is a potential observation or a set of potential observations inconsistent with the hypothesis. Concerning the principle of falsification, the cannonball hypothesis is scientific since any observations where the cannonballs do not hit the ground simultaneously are inconsistent with the hypothesis.

Although falsificationism solves the problem with universal hypotheses, it also has weaknesses. One problem is that interesting hypotheses rarely stand on their own legs. They almost always depend on the occurrence of assumptions and other theories. For example, the hypothesis of the two cannonballs depends on the assumption that cannonballs are heavy, so the effect of air resistance is negligible. The problem with such dependencies is that if an observation is inconsistent with the hypothesis, it is not necessarily the hypothesis that is the culprit. It could also be an assumption we have made or a theory on which we have based ourselves that is incorrect.

This may not seem like a big concern for the hypothesis above. Still, this hypothesis is only a special case of the more general hypothesis that motion under

---

[2] The origin of the legend is a biography published by one of Galilei's assistants. Many historians doubt that this experiment took place [54]. Possibly Galilei has been credited with a similar experiment from a church building in Delft performed by others. However, Galilei did related experiments based on an oblique plane [41].

gravity is independent of mass. To test the general hypothesis for very light objects, we must either remove the air or calculate the effect of air resistance. In both cases, we depend on the surrounding theory. In other words, an inconsistent observation falsifies the hypothesis and its preconditions as a whole and provides little information about what is wrong. Example 3.3 on how the planet Neptune was discovered because Uranus's orbit did not fully comply with Newton's laws of gravity is a good illustration. The deviations in Uranus's orbit could be interpreted as a falsification of Newton's theory, but the cause was an, at that time, undiscovered planet. On the other hand, as described in Example 2.5, the deviation in Mercury's orbit was a real falsification of Newton's theory. The astronomers thought otherwise, however, and were instead searching for the planet Vulcan.

## 14.1.6 Paradigm thinking

As we have seen, falsificationism is not without problems. Thomas Kuhn (1922–1996) was also critical of falsificationism, but his criticism was, first and foremost, the mismatch between falsificationism and historical realities [6]. According to classical falsificationism [14], the progress of science may be summarized along the following lines: Science is based on problems. Researchers propose falsifiable hypotheses as to how these problems can be solved. The proposed hypotheses are then critically evaluated. Some hypotheses will be quickly eliminated, while others will survive the initial phase. The latter will be subject to even more critical evaluations. When a hypothesis that has resisted a wide range of detailed tests is eventually falsified, a new problem has arisen that hopefully goes far beyond the original, now solved problem. The new problem gives rise to new hypotheses, which in turn are evaluated, and so on. This way, it goes on forever. It can never be said of a theory that it is true, but hopefully, it is better than its predecessor as it has withstood more tests.

Unfortunately, this is not in line with how science has evolved historically. In fact, if the scientists of the past had based themselves on falsificationism, many of our most fruitful theories would never have appeared. As we have seen, the deviation in Uranus's orbit did not lead to the rejection of Newton's theory. Instead, it initiated a search that led to the discovery of Neptune. The deviation in Mercury's trajectory was first explained by Einstein's theory of relativity.

Kuhn's view of the evolution of science is illustrated in Figure 14.2. A paradigm is a regime within which research is conducted. A paradigm includes a theory and methodology for testing and evaluation. The theory is not necessarily without problems and anomalies but is sufficiently solid to enable fruitful research. A paradigm can work well over a long period of time. Newton's theory of gravity is an excellent example in that respect. It enabled great advances in physics for hundreds of years. Within a paradigm, so-called normal science is conducted. As normal science progresses, we reach a point where it becomes challenging to continue within the same

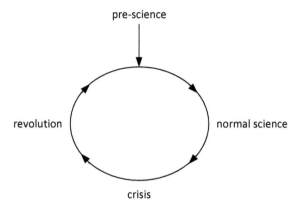

pre-science

revolution                                    normal science

crisis

**Fig. 14.2** Kuhn's view on the evolution of science.

paradigm. Science is in crisis. This crisis leads to a scientific revolution giving rise to a new paradigm. Within the new paradigm, scientists will pursue normal science until the next crisis, and so on. The introduction of Einstein's theory of relativity that revolutionized physics is an example of such a paradigm shift. The first half of the twentieth century was perhaps the most fertile period of physics research ever.

One problem with Kuhn's model for the evolution of science is representing paradigm-independent progress. Should we argue that a paradigm is better than any other paradigm, there must be paradigm-independent criteria that they can be measured by. Kuhn's publications are unclear on this. His publications indicate that Kuhn initially favored a relativistic view, namely that the quality of a paradigm can only be measured in the context of a paradigm. Later he spent much time distancing himself from such an interpretation.

Another example of paradigm thinking is Imre Lakatos's (1922–1974) so-called research programs [61]. A research program is roughly a paradigm within which the various components are weighted in importance or significance. The weighting represents that specific laws and principles are more trustworthy or fundamental than others. Lakatos emphasized the existence of a hard core as the defining characteristic of a research program. Unlike Kuhn, Lakatos was a proponent of falsificationism. Because different parts of a research program have different weighting, it is among the more peripheral and less essential elements of the research program that we look for errors when a hypothesis is falsified. That way, it is not a theory in its entirety that is falsified, as Popper's critics argue, but the parts of the theory we have the least faith in. Lakatos saw his philosophy of science as a tool for science historians and not as a methodology for ongoing research – in other words, as a tool for retrospectively explaining developments in a science.

### 14.1.7 Epistemological anarchism

Paul Feyerabend (1924–1994) argued that there are no useful rules that, without exception, control the progression of science or accumulation of knowledge. The history of science is complex, and if we insist on the existence of a general methodology to be followed slavishly without hindering scientific progress, it will consist of only one rule, namely the useless statement *everything is allowed*. This he called epistemological anarchism [65].

According to Feyerabend, successful researchers are methodological opportunists who take any liberty as long as it is useful. Feyerabend illustrated his position by showing how Galileo Galilei used rhetoric, propaganda, and other "dirty" tricks to argue for the heliocentric theory of the solar system.

Feyerabend promoted a relativistic view of the history of science. Sciences evolve, and new paradigms emerge, but we cannot claim that one paradigm is better than any other, as this presupposes the existence of a paradigm-independent measure. According to Feyerabend, science is just one of many forms of thinking that man has developed, and not necessarily the best. Science is a collection of theories, practices, research traditions, and views whose application domain is unclear and whose utility vary.

### 14.1.8 Probabilism

As mentioned in Section 14.1.6, Lakatos suggested using importance or significance to measure or weight the constituents of a paradigm. The weighting should represent that certain principles and laws are more fundamental than others. Weighting can be understood as a measure of trust – a belief that will vary over time. Belief in the theory of the Higgs particle increased substantially when in 2013, CERN published empirical evidence for its existence. Around 1850 the confidence in Newton's theory of gravity was almost absolute. Today, we still have high confidence in Newton's theory, but only for a limited domain.

Trust is often measured as a probability. It has given rise to the direction of philosophy known as probabilism. There are two main variants [83], namely:

- *Objective probabilism:* Probabilities represent what a rational agent may deduce based on available knowledge.
- *Subjective probabilism:* Probabilities represent subjective confidence.

There are different sub-variants of both. They differ concerning the assignment and representation of trust as well as the rules for reasoning and deduction.

Objective probabilism is restrictive, while subjective probabilism is liberal. A subjective entity may believe that the Earth is flat. Therefore, the statement "the Earth is flat" may be assigned high probability in subjective probabilism. However, this assignment cannot be justified rationally and is therefore wrong from an objective point of view.

### 14.1.9 Experimentalism

Experimentalism is a collective term for that philosophy of science that emphasizes the value of experiments and questions the usefulness of theory. Ian Hacking (b. 1936) and his experimental realism [33] is a well-known representative of this direction. Experimentalism builds on the existence of experiments independent of large-scale theories. The experimentalists emphasize that we have a wide range of practical strategies that enable the establishment of experimental facts and do not depend on theories to any great extent. At the extreme, experimentalists see scientific progress as a gradual accumulation of experimental knowledge, knowledge not threatened by scientific revolutions.

Deborah Mayo (b. 1951) argues [55] that experiments have a life independent of large-scale theories. She emphasizes that:

- Many experiments have other motives than testing or confirmation of a theory. In technology science, for example, our main concern is whether the artifact needs have been fulfilled.
- Experimental data can be justified independently of theory. An experiment performed by competent researchers has a well-defined layout and is documented so that others can repeat and check the findings. An experiment is set up with given parameters and observes the effects. These observations are theory-independent even if their interpretation is not.
- Experimental knowledge exists even though theories change. Suppose we did an experiment to test a particular theory. If the theory changes, maybe the experiment is less relevant, but the layout still "exists" provided it is appropriately documented. The same holds for the results of the experiment – that is, the observations made for the various stimuli. What may not survive is the interpretation of these results.

In short, according to experimentalism, experimental data is the hard core of science – the knowledge that survives from one scientific paradigm to the next.

## 14.2 Technology science in this picture

Given the role of technology science as a driving force in modern society, there is surprisingly little literature on the philosophy of technology science. One reason might be that technology science has been perceived as a branch of natural science that does not provide philosophical problems on its own. On the other hand, technology has attracted quite some interest from philosophers, especially in recent years. Some directions are:

- technological knowledge compared to other kinds of knowledge;
- technology and human nature;
- technology as a characteristic of modern society.

See, for example, [59], [72].

This book is, however, limited to technology science. In the following, we relate technology science as defined and presented in this book to the philosophical directions outlined above.

## 14.2.1 Technology science and empiricism

Technology science addresses human needs. Capturing these needs and assessing their degree of fulfillment requires empirical studies. Hence, technology science generates a posteriori knowledge. Technology science depends on logic, mathematics, and statistics – knowledge that may be perceived as a priori. Whether or not this knowledge is a priori is for a technology scientist, and the contributions of this book, of little relevance.

## 14.2.2 Technology science and inductionism

From the time of Bacon until the beginning of the twentieth century (and Einstein's theory of relativity), the prevailing mindset was inductionistic. It was believed that reliable knowledge could be obtained by induction [27]. A typical inductionistic view on the practice of science was as follows:

- *Step 1:* Observations are collected without any selection strategy or guess about importance.
- *Step 2:* The observations are analyzed, compared, and classified without other hypotheses or postulates than common sense.
- *Step 3:* Generalizations are made from the analyzed and classified observations inductively.
- *Step 4:* From then on, research is both deductive and inductive. Deductions may be made from generalizations, and generalizations may be made based on deductions.

If this book had advocated an inductionistic view as outlined above, Chapter 6 would have described a process following the four steps above. However, Chapter 6 does not, and there are good reasons for that.

A significant problem with Step 1 is that there are infinitely many possible observations. Collecting observations without a working hypothesis about what observations are relevant is inefficient.

We encounter a similar problem with Step 2. It is not practical to analyze, compare and classify observations without an initial idea or hypothesis about what we look for. There are infinitely many classifications and relations between observations.

As for Step 3, the problem is how we arrive from a finite number of observations at a statement about an infinite or a substantial population. Put differently, even if the same experiment produces the same result a million times in a row, there is no guarantee that it will always do so. Step 4 also involves inductive generalization and suffers from the same problem.

Rejecting induction as a research method is not incompatible with the human brain implicitly using principles similar to induction to solve problems. For example, the British empiricist David Hume (1711–1776) distanced himself from the principle of induction but at the same time accepted induction as a psychological mechanism that we can hardly survive without [36].

The practical relevance of induction depends on what is understood by induction. Our outline above touches on only one aspect of this complex discussion.

### 14.2.3 Technology science and positivism

Technology science depends on physics as well as chemistry. Laser technology would have been impossible without quantum physics, and modern pharmaceutics builds on chemistry. A genetically modified tomato is also a kind of artifact. Its design is entirely dependent on biology. Technology science is tightly interwoven with sociology since technology has to meet human needs and support social processes and structures. We can therefore argue that technology science is at least as complex as sociology. Whether it is more complex is debatable. Since sociology also depends on technology science, the best fit for Comte's classification is to view technology science as an ingredient of sociology.

Figure 14.3 illustrates that technology science, like explanation science, can be understood as cross-cutting disciplines in Comte's hierarchy.

Comte argued that the classification of the sciences is a prerequisite for a theory of technology. More complex phenomena are (complex in Comte's meaning, see Figure 14.3), easier to modify. High complexity makes it easier to replace a natural order with an artificial human order, that is, to introduce an artifact [8].

Technology science is compatible with positivism in its emphasis on the scientific method. However, Comte's natural-scientific attitude towards understanding man and society is problematic from a technology science perspective. As explained previously in the book, action research combined with qualitative studies based on in-depth interviews is often the best practice for evaluating artifacts in realistic con-

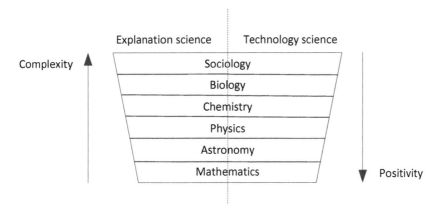

**Fig. 14.3** Technology science in Comte's hierarchy of sciences.

texts. A purely natural-scientific approach fits poorly in such settings as there are too many complex parameters to control.

### 14.2.4 Technology science and logical empiricism

Technology science cannot be reconciled with a restrictive interpretation of the demarcation principle. The human needs that a new technology is supposed to fulfill will lead to universal hypotheses over infinitely large domains, and such statements cannot be verified once and for all. However, separating the scientific from the non-scientific is also vital for technology science. Technology science, like other sciences, needs to distinguish the scientifically acceptable from the unacceptable. Most new artifacts make use of already existing technology. This technology must be appropriately documented. Proper documentation is also necessary to convince serious stakeholders that our assertions and findings are trustworthy.

### 14.2.5 Technology science and falsificationism

Falsification is a crucial principle of technology science, even though it is not necessarily referred to as such. Testing is a good example. Chapters 7–10 build on a falsificationist mindset. The term falsificationism is also used more widely as an explanatory model for the evolution of science. In this book, however, we are primarily concerned with the research process, from identifying artifact needs to evaluating and publishing. How technology science has evolved historically and will evolve in the future is another issue.

Falsificationism has been criticized because the failure of a prediction does not necessarily falsify the hypothesis. It might be an assumption or the underlying theory that is wrong. In practice, this usually is not a problem. Another issue is how technology science has evolved historically and will evolve in the future.

### 14.2.6  Technology science and paradigm thinking

As pointed out above, this book does not attempt to explain the evolution or history of science. Technology science methodology, as described by us, is compatible with falsificationism as well as Kuhn's or Lakatos's paradigm thinking. Concerning Kuhn's position on the development of sciences, we mainly address the phase of normal science.

### 14.2.7  Technology science and epistemological anarchism

Like other sciences, technology science is not a homogeneous, conflict-free structure. Technology scientists differ significantly in how they conduct research. The advice, recommendations, and guidelines put forward in this book are not prerequisites for success. They may, however, increase the chances of success.

Feyerabend's view on the lack of paradigm-independent progress is hard to swallow. Much of the technology that is common today was utterly unthinkable a few decades ago. Moreover, I can hardly imagine what kind of technology we could have built, for example, in the eighteenth century that we cannot make today. To claim that this does not represent progress seems unreasonable.

I accept that there are many approaches to technology science and that it might not be a good idea to impose restrictions on the researchers during the invention phase. However, this does not mean that we cannot distinguish technology science from non-science.

What sets technology science apart from non-science is the invention of new or improved artifacts that are properly evaluated and documented. In some disciplines of technology science, the expectations for validity are as for natural science. In others, concerned with the behavior of humans and social structures, validity is more problematic.

### 14.2.8 Technology science and probabilism

Much technology is tangible. We can use or test it ourselves. We can, for example, touch and smell a car engine and personally observe how it thrusts the car forward. A probabilistic mindset may therefore seem unnecessary.

On the other hand, probabilistic reasoning is commonly required to develop a technology. The reliability of technology can, for instance, be measured probabilistically. Does this imply that technology science builds on probabilism?

The question is not trivial. However, we must distinguish between a probabilistic hypothesis and a probabilistic characterization of confidence in a hypothesis. For a fair hexagonal dice, we can assert: The probability that we get 1, 2, or 3 when throwing the die is 0.5. Our confidence in this hypothesis is probability 1. Probabilism, as a philosophical direction, concerns the latter probability, probability as a measure of confidence.

### 14.2.9 Technology science and experimentalism

Technology science is hard to reconcile with experimentalism. Many technologies build on theories that go way beyond what is available as theory-independent, accumulated, experimental knowledge. Nuclear power plants require reliable storage of waste for thousands of years. We cannot address such needs without theories on Earth's geological evolution. The same holds for space research. Building satellites for the successful exploration of alien celestial bodies presupposes large-scale theories.

As with other sciences, a theory is a facilitator for new inventions. Quantum computers [42] would not exist without quantum theory.

An alternative to accumulated, experimental knowledge as paradigm-independent knowledge could be accumulated technology. For example, we could envision measuring the value of a paradigm by the technology the paradigm gave rise to.

# Appendix A
# Overview of Definitions

**A posteriori knowledge**  *A posteriori knowledge* is knowledge that is based on experience.

**A priori knowledge**  *A priori knowledge* is knowledge that is not based on experience.

**Alternative hypothesis**  An *alternative hypothesis* characterizes the alternative to the null hypothesis, the alternative we are trying to argue for.

**Anonymized data**  *Anonymized data* is data where personal identity cannot be derived directly, indirectly, or via a scrambling key.

**Applied research**  *Applied research* is research to find solutions to a practical human problem.

**Artifact**  An *artifact* is a thing, an object, or a phenomenon created by humans.

**Basic research**  *Basic research* is research aimed at satisfying the need to know.

**Causal relation**  A *relationship* between two events A and B is *causal* if (1) there is a correlation between A and B in the sense that B occurs every time A occurs; (2) A always occurs before B in time – that is, A and B are strictly arranged in time; (3) there is no plausible alternative explanation for the correlation between A and B.

**Cognition**  *Cognition* is the activity of perceiving something as it is (in reality), regardless of the cognitive subject.

**Computer simulation**  *Computer simulation* involves simulating a system or artifact using software.

**Conclusion valid evaluation**  An *evaluation* is *conclusion valid* if the conclusions drawn are correct given the observations made.

**Construct valid evaluation**  An *evaluation* is *construct valid* if abstract concepts, constructs, and relationships from surrounding theories, hypotheses, and assumptions are correctly operationalised in the form of measurement points, indicators, and metrics.

K. Stølen, *Technology Research Explained*, https://doi.org/10.1007/978-3-031-25817-6

**De-identified data**  *De-identified data* is data where identification of individuals is made difficult by representing personal identity with a scrambling key whose link key is stored separately.

**Directly identifiable data**  *Directly identifiable data* is data where the identity of individuals is clearly stated.

**Duplicate publication**  *Duplicate publication* is the reuse of one's own already published material in such a way that this material in the new publication appears new and original.

**Empirical hypothesis**  An *empirical hypothesis* is a hypothesis about reality.

**Existential hypothesis**  An *existential hypothesis* asserts that a specific population contains at least one specimen with a certain characteristic.

**Experimental simulation**  An *experimental simulation* is an experiment that simulates a relevant part of reality under controlled conditions.

**Explanation science**  *Explanation science* is science aiming at understanding reality as it is.

**Externally valid evaluation**  An *evaluation* is *externally valid* if the domain, the situation, the context, and the time aspect for which it is alleged to apply or be relevant are correctly characterized.

**Fact**  A *fact* is a true statement about some past event.

**Field experiment**  A *field experiment* is an experiment conducted in a natural environment, but where the researcher intervenes and manipulates certain factors.

**Field study**  A *field study* involves direct observation of a system, with the least possible interference from the researcher.

**Hypothesis**  A *hypothesis* is an educated guess expressed as an assertion.

**Impact goal**  An *impact goal* describes why the project has been established, for example, helping to reach a desired future societal state (in many cases, not within the lifetime of the project).

**Implication**  An assertion is an *implication* of (or is implied by) another assertion if it (the former assertion) follows with necessity (from the latter assertion).

**Implication requirement for existential hypotheses**  A prediction satisfies the *implication requirement for existential hypotheses* if the hypothesis is an implication of the prediction, the assumptions, and the facts together (the conjunction of the prediction, the assumptions, and the facts).

**Implication requirement for statistical hypotheses**  A prediction satisfies the *implication requirement for statistical hypotheses* if the null hypothesis, the assumptions, and the facts together imply that the probability of the prediction being "wrongly" falsified is less than or equal to the significance level.

**Implication requirement for universal hypotheses**  A prediction satisfies the *implication requirement for universal hypotheses* if it is an implication of the hypothesis, the assumptions, and the facts together (the conjunction of the hypothesis, the assumptions, and the facts).

**Implicit hypothesis**  An *implicit hypothesis* is a hypothesis that follows implicitly or can be deduced from other available documentation.

**In-depth interview**  An *in-depth interview* involves structured collection of a (usually large) amount of information from relatively few individuals.

**Initial hypothesis**  An *initial hypothesis* is a hypothesis that can be deduced from the problem analysis (that is, the documentation established as part of the problem analysis).

**Innovation**  *Innovation* is the production or adoption, assimilation, and exploitation of a value-added novelty in economic and social spheres; renewal and enlargement of products, services, or markets; development of new methods of production; or the establishment of new management systems. It is both a process and an outcome.

**Inter-observer reliable evaluation**  An *evaluation* is *inter-observer reliable* if the observations (assessments/estimations) in each observer group conform with each other.

**Internal consistency reliable evaluation**  An *evaluation* is *internal consistency reliable* if there is a natural mutual correlation between the results from different value-setting elements and between value-setting elements and the overall outcome of the evaluation.

**Internally valid evaluation**  An *evaluation* is *internally valid* if the causal relations that the evaluation claims to have established are real.

**Knowledge**  *Knowledge* is what is learnt or understood, what we know.

**Laboratory experiment**  A *laboratory experiment* is an experiment where the researcher has considerable control and ability to isolate the investigated variables.

**Logic**  *Logic* is non-empirical reasoning based on sound rules of deduction and argumentation.

**Mathematics**  *Mathematics* is a non-empirical study of abstract structures, their properties, and patterns.

**Method-oriented hypothesis**  A *method-oriented hypothesis* is a hypothesis that makes some sort of claim about which research method is suitable for solving a particular problem.

**Method triangulation**  *Method triangulation* implies that certain phenomena are studied from several angles or points of view utilizing different methods.

**Negation**  A claim is a *negation* of another claim if the former expresses the exact opposite of the latter.

**Null hypothesis**  A *null hypothesis* characterizes the null state, for example, the current situation, the prevailing perception, or what is possible with already existing solutions or technology.

**Outcome goal** An *outcome goal* describes what a project or measure should achieve and is linked to the direct results and outputs of the project.

**Parallel-forms reliable evaluation** An *evaluation* is *parallel-forms reliable* if the choice between variants of the same evaluation method has almost no impact on the result.

**Patent** A *patent* is an exclusive right to a product or a process that generally provides a new way of doing something or offers a new technical solution to a problem. To get a patent, technical information about the invention must be disclosed to the public in a patent application.

**Prediction** A *prediction* is an assertion about a future condition or circumstance.

**Primary data** *Primary data* is data collected or generated as part of the project.

**Privacy** *Privacy* is the ability of an individual or group to seclude themselves or information about themselves and thereby express themselves selectively.

**Processed data** *Processed data* is data that has emerged through interpretation and processing.

**Prototyping** *Prototyping* involves building a model of an artifact.

**Raw data** *Raw data* is original, untreated data.

**Reality** *Reality* is the outer world. What we face in our natural attitude to the outside world, and what we can indirectly deduce about it.

**Reliable evaluation** An *evaluation* is reliable if it can be repeated and gives approximately the same result each time it is repeated.

**Repeatability** *Repeatability* means that other researchers should be able to repeat our studies and experiments.

**Research** *Research* is a systematic process for generating new knowledge.

**Research method** A *research method* is a specialized approach or procedure to conduct research.

**Science** *Science* is a systematically arranged body of methodologically established knowledge.

**Scientific article** A *scientific article* is an article that presents new, not previously published knowledge documented in a testable manner.

**Secondary data** *Secondary data* is data used by the project but with external origin.

**Solution-oriented hypothesis** A *solution-oriented hypothesis* is a hypothesis that claims more than what follows implicitly from the problem analysis.

**Stakeholder** A *stakeholder* is one participating in or having interests in some enterprise.

**Statistical hypothesis** A *statistical hypothesis* is an assertion about the value of one (or more) parameter(s) for a specific population. It can claim something about the parameter's value (such as average, size, or median) or its probability distribution (such as normal, uniform, or logarithmic distribution).

**Success criterion** A *success criterion* characterizes a test or condition that must be fulfilled to succeed with a given enterprise.

**Survey** A *survey* is a collection of information from a wide and carefully selected variety of stakeholders.

**Technology** *Technology* includes all human-made objects and the skills we use to manufacture and employ them.

**Technology science** *Technology science* is science aiming at expanding reality with new or significantly better artifacts.

**Test-retest reliable evaluation** An *evaluation* is *test-retest reliable* if the result remains approximately the same independent of when it is repeated.

**Testability** *Testability* means that other researchers should be able to check our results.

**Theory** A *theory* is a system of (partially) confirmed statements that determine or explain the relations between phenomena.

**Universal hypothesis** A *universal hypothesis* asserts that every member of a specific population has a particular characteristic.

**Valid evaluation** An *evaluation* is *valid* if it evaluates what it is meant to evaluate.

**Working hypothesis** A *working hypothesis* is a preliminary hypothesis we accept as a basis for further research.

# References

1. Aitkenhead, D.: Peter Higgs: I wouldn't be productive enough for today's academic system. The Guardian, 6.12 (2013)
2. Anglia Ruskin University: Harvard system of referencing guide. Accessed 22.8 2017: https://libweb.anglia.ac.uk/referencing/harvard.htm
3. Bain, R.: Technology and state government. American Sociological Review **2**, 860–874 (1937)
4. Baskerville, R.: Investigating information systems with action research. Communications of the Association for Information Systems **2** (1999). Article 19
5. Bhattacharyya, G., Johnson, R.: Statistical Concepts and Methods. Wiley (1977)
6. Bird, A.: Thomas Kuhn. The Stanford Encyclopedia of Philosophy (Fall 2013 Edition): https://plato.stanford.edu/archives/fall2013/entries/thomas-kuhn/
7. Blades, R.: AI generates hypotheses human scientists have not thought of. Scientific American, 28.10 (2021)
8. Bourdeau, M.: Auguste Comte. The Stanford Encyclopedia of Philosophy (Summer 2011 Edition): https://plato.stanford.edu/archives/sum2011/entries/comte/
9. Brown, J., Fehige, Y.: Thought experiments. The Stanford Encyclopedia of Philosophy (Summer 2017 Edition): https://plato.stanford.edu/archives/sum2017/entries/thought-experiment/
10. Callander, B.: The critical twist. Air Force Magazine **72**, 150–156 (1989)
11. Cambridge English Dictionary: Educated guess. Accessed 2.4 2021: http://dictionary.cambridge.org/dictionary/english/educated-guess
12. CBSNews/AP: Quantas crew faced 54 alarms warning of failure. 18.11 (2010)
13. CERN Press Office: New results indicate that particle discovered at CERN is a Higgs boson. 14.3 (2013)
14. Chalmers, A.: What Is This Thing Called Science? Open University Press (1999)
15. Crang, M., Cook, I.: Doing Ethnographies. SAGE (2007)
16. Creath, R.: Logical empiricism. The Stanford Encyclopedia of Philosophy (Fall 2017 Edition): https://plato.stanford.edu/archives/fall2017/entries/logical-empiricism/
17. Creswell, J.: Qualitative Inquiry and Research Design: Choosing Among Five Traditions. SAGE (2007)
18. Cross, N.: Designerly ways of knowing: Design dicipline versus design science. Design Issues **17**, 49–55 (2001)
19. Crossan, M., Apaydin, M.: A multi-dimensional framework of organizational innovation: A systematic review of the literature. Journal of Management Studies **47**, 1154–1191 (2010)
20. Darwin, C.: On the Various Contrivances by Which British and Foreign Orchids are Fertilised by Insects. John Murray (1862)

21. Dresch, A., Lacerda, D., Antunes, J.: Design Science Research. Springer (2015)
22. Edvards, B.: The space elevator – NIAC phase II final report. Tech. rep., Eureka Scientific (2003)
23. Encyclopaedia Britannica: Abominable snowman. Accessed 08.4 2021: https://www.br itannica.com/topic/Abominable-Snowman
24. Fagerberg, J.: A guide to Schumpeter. In: C. Mitcham, R. Mackey (eds.) Confluence: Interdisciplinary Communications, pp. 20–22. Centre for Advanced Study, Oslo (2009)
25. FAI portal: 11 October 2005: Russian cosmonaut Krikalev becomes the absolute record holder in accumulated space flight time. 10.10 (2015)
26. Falk, D.: How artificial intelligence is changing science. Quanta Magazine, 11.3 (2019)
27. Føllesdal, D., Walløe, L.: Argumentasjonsteori og Vitenskapsfilosofi. Universitetsforlaget (1977)
28. Friedel, R., Israel, P., Finn, B.: Edison's Electric Light: Biography of an Invention. Rutgers University Press (1986)
29. Garcia-Ceja, E., Hugo, Å., Morin, B., Hansen, P.O., Martinsen, E., Lam, A.N., Haugen, Ø.: Towards the automation of a chemical sulphonation process with machine learning. In: Proceedings of the 7th International Conference on Control, Mechatronics and Automation, pp. 352–357. IEEE (2019)
30. Gillies, D., Gillies, M.: Artificial intelligence and philosophy of science from the 1990s to 2020. In: Proceedings of the 25th Conference on Contemporary Philosophy and Methodology of Science. University of A Coruña (2020)
31. Grant, M., Booth, A.: A typology of reviews: An analysis of 14 review types and associated methodologies. Health Information and Libraries Journal 26, 91–108 (2009)
32. Gribbin, J.: Science: A History. Allen Lane (2002)
33. Hacking, I.: Experimentation and scientific realism. Philosophical Topics 13, 71–87 (1982)
34. Hamlyn, D.: Empiricism. In: D. Borchert (ed.) Encyclopedia of Philosophy (2nd edition), vol. 3, pp. 213–221. Macmillan (2006)
35. Hawking, S.: How to build a time machine. Daily Mail Online, 27.4 (2010)
36. Henderson, L.: The problem of induction. The Stanford Encyclopedia of Philosophy (Spring 2019 Edition): https://plato.stanford.edu/archives/spr2019/entries/induction-problem/
37. Henderson, R., Clark, K.: Architectural innovation: The reconfiguration of existing product technologies and the failure of existing firms. Administrative Science Quarterly 35, 9–30 (1990)
38. Hevner, A., Chatterjee, S.: Design Research in Information Systems. Springer (2010)
39. Hevner, A., March, S., Park, J., Ram, S.: Design science in information systems research. MIS Quarterly 28, 75–105 (2004)
40. International Committee of Medical Journal Editors (ICMJE): Recommendations for the conduct, reporting, editing, and publication of scholarly work in medical journals (2016)
41. Johnson, G.: The Ten most Beautiful Experiments. Vintage Books (2008)
42. Kaye, P., Laflamme, R., Mosca, M.: An Introduction to Quantum Computing. Oxford University Press (2007)
43. Kennefick, D.: Testing relativity from the 1919 eclipse – A question of bias. Physics Today 62, 37–42 (2009)
44. Kitchenham, B., Carters, S.: Guidelines for performing systematic literature reviews in software engineering – version 2.3. Tech. Rep. EBSE-2007-01, Keele University (2007)
45. Klein, J.: Francis Bacon. The Stanford Encyclopedia of Philosophy (Winter 2012 Edition): https://plato.stanford.edu/archives/win2012/entries/francis-bacon/
46. van der Kloot, W.: Lawrence Bragg's role in the development of sound-ranging in World War I. Notes and Records of the Royal Society 59, 273–284 (2005)
47. Kritsky, G.: Darwin's Madagascan hawk moth prediction. American Entomologist 37, 206–210 (2001)

48. Lazar, J., Feng, J., Hochheiser, H.: Research Methods in Human-Computer Interaction. Wiley (2010)
49. Levenson, T.: The Hunt for Vulcan – How Albert Einstein Destroyed a Planet and Deciphered the Universe. Head of Zeus (2015)
50. Levy, Y., Ellis, T.: A systems approach to conduct an effective literature review in support of information systems research. Informing Science Journal **9**, 181–212 (2006)
51. Lewin, K.: Frontiers in group dynamics. Human Relations **1**, 5–41 (1947)
52. Lindquist, H.: Historien om Sverige – Storhet och Fall. Norstedts (1995)
53. Lokalhistoriewiki: Peder Pedersen Dagsgardsødegård. Accessed 10.11 2013: https://lo kalhistoriewiki.no/index.php/Vass-Per
54. Martinez, A.: Science Secrets – The Truth About Darwin's Finches, Einstein's Wife, and Other Myths. Pittsburgh University Press (2012)
55. Mayo, D.: The new experimentalism, topical hypotheses, and learning from error. In: Proceedings of the Biennial Meeting of the Philosophy of Science Association, pp. 270–279. University of Chicago Press (1994)
56. McGrath, J.: Groups: Interaction and Performance. Prentice-Hall (1984)
57. Merriam-Webster: Research. Accessed 28.4 2012: https://www.merriam-webster .com/dictionary/research
58. Merriam-Webster: Technology. Accessed 28.4 2012: https://www.merriam-webst er.com/dictionary/technology
59. Mitcham, C., Mackey, R. (eds.): Philosophy and Technology: Readings in the Philosophical Problems of Technology. Collier Macmillan (1983)
60. Mitchell, R.: Action anthropology. Lambda Alpha Journal of Man **2**(2), 40–46 (1970)
61. Musgrave, A., Pigden, C.: Imre Lakatos. The Stanford Encyclopedia of Philosophy (Winter 2016 Edition):https://plato.stanford.edu/archives/win2016/entries/ lakatos/
62. Osborn, A.: Applied Imagination. Scribner's (1953)
63. Popper, K. (ed.): Unended Quest – An Intellectual Autobiography. Routledge (1992)
64. Porter, N. (ed.): Physicists in Conflict. Institute of Physics Pub. (1998)
65. Preston, J.: Paul Feyerabend. The Stanford Encyclopedia of Philosophy (Winter 2012 Edition): https://plato.stanford.edu/archives/win2012/entries/feyera bend/
66. Pretz, K.: Stop calling everything AI, machine-learning pioneer says: Michael I. Jordan explains why today's artificial-intelligence systems aren't actually intelligent. IEEE Spectrum, 31.3 (2021)
67. Radick, G.: Beyond the 'Mendel-Fisher controversy'. Science **350**, 159–160 (2015)
68. Randall, L.: Higgs Discovery – The Power of Empty Space. The Bodley Head (2012)
69. Randall, L.: Knocking on Heaven's Door. Vintage (2012)
70. Roll-Hansen, N.: Ideological obstacles to scientific advice in politics? Tech. Rep. Makt- og demokratiutredningens rapportserie: 48, Samfunnsvitenskapelig fakultet, Universitetet i Oslo (2002)
71. Rothschild, W., Jordan, K.: A revision of the lepidopterous family Sphingidae. Novitates Zoology **9**, supplement (1903)
72. Scharff, R., Dusek, V. (eds.): Philosophy of Technology: The Technological Condition. Blackwell (2003)
73. Schumpeter, J.: The Theory of Economic Development. Harvard University Press (1934)
74. Shadish, W., Cook, T., Campbell, D.: Experimental and Quasi-Experimental Designs for Generalized Causal Inference. Houghton Mifflin (2002)
75. Silver, N.: The Signal and the Noise: The Art and Science of Prediction. Penguin (2013)
76. Simon, H.: The Sciences of the Artificial, 3 edn. MIT Press (1996)
77. Singh, S.: Fermat's last Theorem. Fourth Estate (1997)
78. Solheim, I., Stølen, K.: Technology research explained. Tech. Rep. SINTEF A313, SINTEF IKT (2007)

79. Steffl, A., Cunningham, N., Shinn, A., Durda, D., Stern, S.: A search for Vulcanoids with the STEREO heliospheric imager (2013). ArXiv:1301.3804v1[astro-ph.SR]

80. Stølen, K.: Teknologivitenskap: Forskningsmetode for Teknologer. Universitetsforlaget (2019)

81. Susman, G., Evered, R.: An assessment of the scientific merits of action research. Administrative Sciences Quarterly **23**, 582–603 (1978)

82. Susskind, L.: The Cosmic Landscape. Hachette Book Group (2006)

83. Talbott, W.: Bayesian epistemology. The Stanford Encyclopedia of Philosophy (Winter 2016 Edition):https://plato.stanford.edu/archives/win2016/entries/epis
temology-bayesian/

84. Trist, E., Bamforth, K.: Some social and psychological consequences of the Longwall Method of coal-getting. Human Relations **4**, 3–38 (1951)

85. Uzgalis, W.: John Locke. The Stanford Encyclopedia of Philosophy (Fall 2012 Edition): https://plato.stanford.edu/archives/fall2012/entries/locke/

86. van den Ven, A., Polley, D., Garud, R., Venkataraman, S.: The Innovation Journey. Oxford University Press (2008)

87. Wasserthal, L.: The pollinators of the Malagasy star orchids Angruecum sesguipedule, A. sororium and A. compuctum and the evolution of extremely long spurs by pollinator shift. Botanica Acta **110**, 343–359 (1997)

88. Wieringa, R.: Design Science Methodology for Information Systems and Software Engineering. Springer (2014)

89. Wikipedia: History of aviation. Accessed 8.3 2019: https://en.wikipedia.org/w
iki/History_of_aviation

90. Wikipedia: List of scientific misconduct incidents. Accessed 10.4 2021: https://en.wik
ipedia.org/wiki/List_of_scientific_misconduct_incidents

91. Wikipedia: Privacy. Accessed 21.3 2021: https://en.wikipedia.org/w/index.p
hp?title=Privacy&oldid=1013391969

92. Wikipedia: Synthesis of precious metals. Accessed 30.7 2013: https://en.wikipedia
.org/wiki/Synthesis_of_precious_metals

93. Wikipedia: Willem de Vlamingh. Accessed 7.2 2014: http://en.wikipedia.org/w
iki/Willem_de_Vlamingh

94. Wiktionary: Artifact. Accessed 31.7 2012: http://en.wiktionary.org/w/index.
php?title=artifact&oldid=17093965

95. Wilson, F.: John Stuart Mill. The Stanford Encyclopedia of Philosophy (Spring 2012 Edition): https://plato.stanford.edu/archives/spr2012/entries/mill/

96. Wittemore, R., Knafl, K.: The integrative review: Updated methodology. Journal of Advanced Nursing **52**, 546–553 (2005)

97. World Intellectual Property Organization: What is a patent? Accessed 2.4 2021: https:
//www.wipo.int/patents/en/faq_patents.html

98. Wyllys, R.: Mathematical notes for LIS 397.1: Introduction to research in library and information science. Tech. rep., The University of Texas at Austin School of Information (2007)

99. Yin, R.: Case Study Research: Design and Methods. 3rd edn. SAGE (2003)

# Index

a posteriori knowledge, **155**, 163, *169*
a priori knowledge, **155**, 163, *169*
abstract, 130, 137
    scientific, 130
action research, 25, 52, 115, 143, 164
alternative hypothesis, 12, **106**, *169*
analysis
    problem, 20, 33, 43, 60, 149
    statistical, 2, 52, 121, 143
anonymized data, **57**, *169*
application
    project, 149
applied research, **10**, 130, *169*
architecture
    business, 69
article
    scientific, 21, 41, 47, **131**, 137, *172*
article writing, 22, 41, 132, 137
artifact, v, **8**, 17, 34, 43, 60, 80, 83, 97, 108, 116, 129, 137, 164, *169*
artifact need, 17, 35, 60, 80, 83, 99, 108, 137, 165
assumption, 3, 23, 47, 73, 77, 84, 107, 119, 158, 170

Bacon, Francis (1561–1626), 156, 163
Bain, Read (1892–1980), 9
basic research, **10**, 28, 137, *169*
basic research project, 28
Bessel, Friedrich Wilhelm (1784–1846), 24
biology, 14, 73, 157, 164
Blondlot, Prosper-René (1849–1930), 119
book
    scientific, 47, 134
Bragg, William Lawrence (1890–1971), 10
business, 15, 25, 34, 49, 69, 89, 149

business architecture, 69
business process, 8

causal relation, 53, 78, **117**, *169*
Cayley, George (1773–1857), 13, 67, 68, 100
Chalmers, Alan (b. 1939), 155
characterization
    needs, 18, 35, 40, 137
chemistry, 2, 12, 14, 157, 164
client, 25, 92
cognition, **8**, *169*
compound hypothesis, 70, 111
computer simulation, **51**, 80, 90, 111, *169*
Comte, Auguste (1798–1857), 156, 164
conclusion valid evaluation, 54, **121**, 122, *169*
construct valid evaluation, 54, **119**, 121, *169*
construction process, 3
creativity, 2, 44, 47
criterion, 29
    evaluation, 47
    selection, 53
    success, 19, 23, **39**, 142, *173*

Dagsgardødegård, Peder Pedersen (1782–1846), 10
Darwin, Charles (1809–1882), 76, 98
data
    anonymized, **57**, *169*
    de-identified, **57**, *170*
    directly identifiable, **57**, *170*
    primary, **56**, *172*
    processed, **56**, *172*
    raw, **56**, *172*
    secondary, **56**, *172*
de Fermat, Pierre (1601–1665), 101
de Vlamingh, Willem (1640–1698), 65, 84

de-identified data, **57**, *170*
decision-making process, 40
design
    technology, v, 12, 17, 34, 46, 61, 71, 80, 84,
        110, 120, 139, 164
design process, 2
design science, 2
development
    technology, 28
development project, 28
Digges, Leonard (ca. 1515–ca. 1559), 19
digitalization project, 37
Diophantus (ca. 210–ca. 295), 101
direct evaluation, 99
directly identifiable data, **57**, *170*
dissemination process, 29, 43, 136, 147
doctoral thesis, 133, 139
documentation, 2, 21, 35, 54, 60, 79, 83, 132,
    142
duplicate publication, **135**, *170*

economics, 2
Edison, Thomas Alva (1847–1931), 16, 65, 84,
    87, 110
educated guess, 11, 59, 75, 153
Einstein, Albert (1879–1955), 14, 32, 49, 71,
    120, 160, 163
empirical hypothesis, **12**, 71, *170*
empiricism, 155, 163
    logical, 157, 165
enterprise
    research, 49
epistemological anarchism, 161, 166
evaluation, 12, 18, 34, 50, 59, 77, 83, 97, 105,
    115, 134, 137, 159
    conclusion valid, 54, **121**, 122, *169*
    construct valid, 54, **119**, 121, *169*
    direct, 99
    externally valid, 54, **116**, 121, *170*
    indirect, 68, 99
    inter-observer reliable, **124**, *171*
    internal consistency reliable, **125**, *171*
    internally valid, 54, **117**, 122, *171*
    parallel-forms reliable, **125**, *172*
    reliable, **115**, 138, 150, *172*
    test-retest reliable, **127**, *173*
    valid, **115**, 138, 150, *173*
evaluation criterion, 47
evaluation phase, 19, 34, 50, 59, 83, 101, 137,
    143
evaluation procedure, 64, 83, 97, 105, 111
evaluation process, 47, 111, 148

existential hypotheses
    implication requirement for, **97**, 112, *170*
existential hypothesis, 19, 59, **66**, 72, 76, 83,
    97, 105, 112, *170*
experiment, 19, 48, 75, 84, 98, 109, 116, 124,
    137, 158, 162
    field, **51**, 80, 88, 109, *170*
    laboratory, **51**, 80, 94, 111, *171*
experiment setup, 2, 80, 87, 100, 109
experimental fact, 162
experimental simulation, **51**, 80, 87, 100, 109,
    112, 117, *170*
experimentalism, 162, 167
explanation science, v, **14**, 17, 22, 38, 43, 59,
    65, 71, 84, 164, *170*
explicit hypothesis, 60, 83, 142, 152
externally valid evaluation, 54, **116**, 121, *170*

fact, **11**, 77, 85, 98, 107, 131, 162, *170*
    experimental, 162
falsification, 50, 59, 73, 109, 158, 165
falsificationism, 158, 165
Feyerabend, Paul (1924–1994), 161, 166
field experiment, **51**, 80, 88, 109, *170*
field study, **51**, 80, 89, 108, *170*
form
    reusable, 66, 80, 86, 99, 107
Frege, Gottlob (1848–1925), 157

Galilei, Galileo (1564–1642), 19, 158, 161
Galle, Johann Gottfried (1812–1910), 24
goal
    impact, **35**, *170*
    outcome, **35**, *172*

Hacking, Ian (b. 1936), 162
Harvey, William (1578–1657), 23
Herschel, William (1738–1822), 23
Higgs, Peter (b. 1929), 103, 135, 161
human need, 17, 25, 43, 93, 163
Hume, David (1711–1776), 164
hypothesis, **11**, *170*
    alternative, 12, **106**, *169*
    compound, 70, 111
    empirical, **12**, 71, *170*
    existential, 19, 59, **66**, 72, 76, 83, 97, 105,
        112, *170*
    explicit, 60, 83, 142, 152
    implicit, 31, **60**, 83, 142, *171*
    initial, **60**, *171*
    method-oriented, **62**, *171*
    null, **105**, 110, *171*
    solution-oriented, **60**, *172*

statistical, 64, **68**, 83, 102, 105, *172*
universal, **65**, 71, 77, 83, 99, 102, 112, 158, *173*
working, 31, **62**, 97, 101, 142, 152, 164, *173*
hypothesis testing, 58, 105

impact goal, **35**, *170*
implication, **84**, 97, 113, *170*
implication requirement for existential hypotheses, **97**, 112, *170*
implication requirement for statistical hypotheses, **107**, 112, *170*
implication requirement for universal hypotheses, **85**, 97, 112, *170*
implicit hypothesis, 31, **60**, 83, 142, *171*
in-depth interview, **51**, 63, 79, 94, 108, 164, *171*
indirect evaluation, 68, 99
inductionism, 156, 163
inductionist, 156
inductionistic, 163
industrial project, 28
initial hypothesis, **60**, *171*
innovation, **15**, 43, *171*
inter-observer reliable evaluation, **124**, *171*
internal consistency reliable evaluation, **125**, *171*
internally valid evaluation, 54, **117**, 122, *171*
interview, 37, 56, 135
in-depth, **51**, 63, 79, 94, 108, 164, *171*
invention, 3, 10, 18, 33, 44, 87, 129, 137
invention phase, 18, 35, 43, 53, 59, 94, 101, 137, 143
issue
problem, 10, 20, 33, 46, 61, 131

Jordan, Karl (1861–1959), 76

Karl XII (1682–1718), 11, 75
knowledge, 1, **7**, 20, 34, 43, 77, 93, 131, 155, *171*
a posteriori, **155**, 163, *169*
a priori, **155**, 163, *169*
Krikalev, Sergej K. (b. 1958), 36
Kuhn, Thomas (1922–1996), 159, 166

laboratory experiment, **51**, 80, 94, 111, *171*
Lakatos, Imre (1922–1974), 160, 166
Le Verrier, Urbain (1811–1877), 14, 24, 78
Lee, Sedol (b. 1983), 103
Lipperhey, Hans (1570–1619), 19
Locke, John (1632–1704), 155
logic, 12, 23, **51**, 92, 108, 156, 163, *171*

logical empiricism, 157, 165
logical positivism, 157, 165
Lucretius (99 BCE–55 BCE), 48

maintenance project, 37
master's thesis, 133, 139
mathematics, 12, **51**, 68, 80, 91, 100, 136, 156, *171*
Mayo, Deborah (b. 1951), 162
McGrath, Joseph (1927–2007), 50
medicine, 2, 25, 30, 140
Mendel, Gregor Johann (1822–1884), 119
method
research, 1, **9**, 30, 50, 59, 79, 112, 119, 135, 137, *172*
method triangulation, **53**, *171*
method-oriented hypothesis, **62**, *171*
Mill, John Stuart (1806–1873), 156

natural science, v, 9, 53, 156, 162, 164
need
artifact, 17, 35, 60, 80, 83, 99, 108, 137, 165
human, 17, 25, 43, 93, 163
needs characterization, 18, 35, 40, 137
needs identification phase, 18, 34, 143
negation, **68**, 99, 105, 110, *171*
Newton, Isaac (1643–1727), 14, 23, 65, 78, 159
null hypothesis, **105**, 110, *171*

objective probabilism, 161
one-researcher project, 38, 46
outcome goal, **35**, *172*
overall process, 1, 17

paradigm thinking, 159, 166
parallel-forms reliable evaluation, **125**, *172*
patent, 65, 87, 110, **134**, *172*
phase, 18
evaluation, 19, 34, 50, 59, 83, 101, 137, 143
invention, 18, 35, 43, 53, 59, 94, 101, 137, 143
needs identification, 18, 34, 143
philosophical problem, 157, 163
philosophy, 9, 71, 155
philosophy of science, 155
physics, 2, 10, 12, 14, 136, 157, 164
planning, 26, 34, 43, 83, 134, 140
Popper, Karl (1902–1994), 59, 71, 158, 160
popular scientific publication, 131
positivism, 156, 164
logical, 157, 165
poster
scientific, 130

prediction, 22, 30, **75**, 84, 97, 105, 121, 153, 166, *172*
preprint, 136
primary data, **56**, *172*
privacy, 39, **56**, *172*
probabilism, 161, 167
  objective, 161
  subjective, 161
problem, 1, 8, 19, 20, 24, 27, 33, 44, 62, 88, 109, 120, 130
  philosophical, 157, 163
problem analysis, 20, 33, 43, 60, 149
problem issue, 10, 20, 33, 46, 61, 131
procedure, 1, 9, 25, 30, 37, 40, 44, 55, 63, 83, 99, 105, 141, 156
  evaluation, 64, 83, 97, 105, 111
process, 37, 46, 100, 109, 117, 143, 156
  business, 8
  construction, 3
  decision-making, 40
  design, 2
  dissemination, 29, 43, 136, 147
  evaluation, 47, 111, 148
  overall, 1, 17
  production, 15, 32, 52, 84
  research, 21, 59, 83, 97, 129, 152, 165
  social, 8, 25, 164
  work, 25, 55, 143
  writing, 46, 129, 134, 150
processed data, **56**, *172*
production process, 15, 32, 52, 84
project
  basic research, 28
  development, 28
  digitalization, 37
  industrial, 28
  maintenance, 37
  one-researcher, 38, 46
  research, 1, 18, 28, 34, 43, 61, 83, 102, 135
  space, 35
  sub-, 29, 38, 43, 59
  technology science, 43, 59, 83
project application, 149
prototyping, 13, 17, **51**, 75, 86, 100, 112, 143, *172*
publication
  duplicate, 135
  popular scientific, 131
publishing, 22, 43, 62, 129, 140, 165

qualitative, 2, 54, 108, 123
qualitative study, 164

quality assurance, 115
quantitative, 54, 58, 123

raw data, **56**, *172*
reading, 1, 10, 20, 41, 48, 130, 150
reality, v, **7**, 22, 36, 51, 80, 85, 98, 117, *172*
relation
  causal, 53, 78, **117**, *169*
reliable
  inter-observer, **124**, *171*
  internal consistency, **125**, *171*
  parallel-forms, **125**, *172*
  test-retest, **127**, *173*
reliable evaluation, **115**, 138, 150, *172*
repeatability, 43, **54**, 126, *172*
repeatable, 25, 115
report
  scientific, 1, 21, 38, 47, 132, 138
report writing, 139
requirement, 3, 15, 29, 35, 43, 66, 76, 86, 99, 142
research, 1, **9**, 17, 34, 44, 59, 76, 84, 115, 117, 129, 159, *172*
  action, 25, 52, 115, 143, 164
  applied, **10**, 130, *169*
  basic, **10**, 28, 137, *169*
research enterprise, 49
research method, 1, **9**, 30, 50, 59, 79, 112, 119, 135, 137, 142, *172*
research process, 21, 59, 83, 97, 129, 152, 165
research project, 1, 18, 28, 34, 43, 61, 83, 102, 135
  basic, 28
reusable form, 66, 80, 86, 99, 107
Rothschild, Walter (1868–1937), 76

Schumpeter, Joseph (1883–1950), 15
science, v, **9**, 23, 48, 61, 75, 83, 120, 129, 138, 156, *172*
  design, 2
  explanation, v, **14**, 17, 22, 38, 43, 59, 65, 71, 84, 164, *170*
  natural, v, 9, 53, 156, 162
  philosophy of, 155
  social, v, 14, 22, 100, 115, 143
  technology, v, 3, **12**, 17, 43, 59, 61, 80, 81, 93, 102, 130, 137, 142, 155, *173*
scientific abstract, 130
scientific article, 21, 41, 47, **131**, 137, *172*
scientific book, 47, 134
scientific poster, 130
scientific practice, 163
scientific report, 1, 21, 38, 47, 132, 138

secondary data, **56**, *172*
selection criterion, 53
self-plagiarism, 129, 135
Semmelweis, Ignaz (1818–1865), 30
setup
  experiment, 2, 80, 87, 100, 109
Simon, Herbert (1916–2001), 2
simulation
  computer, **51**, 80, 90, 111, *169*
  experimental, **51**, 80, 87, 100, 109, 112, 117, *170*
social process, 8, 25, 164
social science, v, 14, 22, 100, 115, 143
social structure, 8, 25, 89, 164
sociology, 2, 12, 156, 164
solution-oriented hypothesis, **60**, *172*
space project, 35
stakeholder, 18, 29, **34**, 49, 135, 165, *172*
statistical analysis, 2, 52, 121, 143
statistical hypotheses
  implication requirement for, **107**, 112, *170*
statistical hypothesis, 64, **68**, 83, 102, 105, *172*
structure
  social, 8, 25, 89, 164
study
  field, **51**, 80, 89, 108, *170*
  qualitative, 164
sub-project, 29, 38, 43, 59
subjective probabilism, 161
success criterion, 19, 23, **39**, 142, *173*
survey, **51**, 63, 80, 93, 111, 117, 123, 143, *173*

Taleb, Nassim Nicholas (b. 1960), 65
Tax, Sol (1907–1995), 28
technology, **9**, 13, 20, 34, 43, 77, 88, 124, 131, 137, *173*
technology design, v, 12, 17, 34, 46, 61, 71, 80, 84, 110, 120, 139, 164
technology development, 28
technology science, v, 3, **12**, 17, 43, 59, 61, 80, 81, 93, 102, 130, 137, 142, 155, *173*
technology science project, 43, 59, 83

test-retest reliable evaluation, **127**, *173*
testability, 21, 29, 43, **54**, 115, 126, 131, 138, 157, *173*
testing
  hypothesis, 58, 105
theory, **11**, 59, 65, 77, 84, 102, 106, 119, 158, *173*
thesis
  doctoral, 133, 139
  master's, 133, 139
thinking
  paradigm, 159
triangulation
  method, **53**, *171*
Tucker, William Sansome (1877–1955), 10

universal hypotheses
  implication requirement for, **85**, 97, 112, *170*
universal hypothesis, **65**, 71, 77, 83, 99, 102, 112, 158, *173*

valid
  conclusion, 54, **121**, 122, *169*
  construct, 54, **119**, 121, *169*
  externally, 54, **116**, 121, *170*
  internally, 54, **117**, 122, *171*
valid evaluation, **115**, 138, 150, *173*
verification, 73

Wiles, Andrew (b. 1953), 101
Wittgenstein, Ludwig (1889–1951), 157
work process, 25, 55, 143
working hypothesis, 31, **62**, 97, 101, 142, 152, 164, *173*
Wright, Orville (1871–1948), 13, 67, 100
Wright, Wilbur (1867–1912), 13, 44, 67, 100
writing, 1, 21, 38, 45, 129, 137, 140
  article, 22, 41, 132, 137
  report, 139
writing process, 46, 129, 134, 150